Question Bank in Fruit Science

About the author

Sambhaji Dattatray Jarande has completed his Ph. D (Horticulture in 2015) and M. Sc (Horticulture in 2011) in Fruit Science with First class from ASPEE College of Horticulture and Forestry Navsari Agricultural University, Navsari (Gujarat) and B.Sc (Horticulture in 2008) from Mahatma Phule Krishi Vidyapeeth, Rahuri (MS). He has been awarded PG fellowship during Ph. D degree sponsored by Navsari Agricultural University, Navsari based on grade point obtained from M.Sc. degree. He has qualified ICAR, NET/ASRB during 2015. He has experience regarding 3.8 years as a Research and Development and extension work. Now, he joined as an Assistant Professor at College of Horticulture, Malegaon, Dist. Nashik. He has published more than 11 research papers, and more than 5 articles on reputed national agricultural journals also. He has published more than eight abstracts and one book, as a co-author during his Ph.D degree under title "Good Agricultural Practices in Mango".

Question Bank in Fruit Science

By

Sambhaji Dattatray Jarande

ASPEE College of Horticulture & Forestry
Navsari Agricultural University
Navsari-396450, Gujarat

books for everyone
PAPERBACKS

nipa books for everyone
PAPERBACKS

A Paperback Division of
NEW INDIA PUBLISHING AGENCY
101, Vikas Surya Plaza, CU Block, LSC Market
Pitam Pura, New Delhi 110 034, India
Phone: + 91 (11) 27 34 17 17 Fax: + 91 (11) 27 34 16 16
Email: info@nipabooks.com
Web: www.nipabooks.com

Feedback at feedbacks@nipabooks.com

ISBN 978-93-85516-79-5

Composed, Designed & Printed in India

Preface

Horticulture is one of the important sectors of agriculture. As per the demand of the present competitive era, it is very important that student should have proper study material for various competitive examinations such as ICAR's JRF, SRF, ARS/NET and other state level examinations. The present book "Question Bank in Fruit Science" is written in very simple and understandable way and includes 3099 multiple choice of questions and fill in the blanks-200 and 25 sample questions, which will help all the aspirants preparing competitive examination. With the help of this book students of the horticulture sector can score more marks and get higher rank and it help them to reach to the desired goal. Presently when we start the preparation of competitive examinations, generally we don't know what to read and what not to read. In the things competitive era it is not all about the mugging up of all the things. So, I thought it will be better if the students get the previous year's question papers and some related matter. By reading this book students can get the idea about how to prepare for the competitive examinations. Besides this experience of the many student of the horticulture field are also included to make this book as a complete guide for competitive examinations.

The present book entitled" Question Bank in Fruit Science" will fulfil a long felt need of Indian students. This volume has been designed to cover the courses offered by various Indian universities for Graduate and Post graduate level and also the ICAR syllabus of various competitive examinations. Hope this book would be useful for graduate and postgraduate students of Indian agriculture students.

To minimise the errors in the book, I tried my level best, however, due to human errors or printing mistakes, there might be some errors. Therefore, readers are always requested to make their valuable suggestions for the further improvement of this book. For any opinion or suggestion please send the email. On: sambhajijarande55@gmail.com

I wish to express my sincerely heartfelt gratitude to my parents, all the scientists and students of Discipline of Horticulture, Navsari Agricultural University, Navsari for their immense help, full co-operation and constant encouragement extended during the preparation of the manuscript for this book.

Dr. Sambhaji Dattatray Jarande
Ph.D (Horticulture) Fruit Science
Aspee College of Horticulture & Forestry
Navsari Agricultural University
Navsari- 396 450.

Contents

1

Multiple Choice of Questions

1. Which fruit is rich in citric acid?
   (a) Apple
   (b) Peach
   (c) Pear
   (d) Plum
2. Pineapple is which kind of fruit?
   (a) Sorosis
   (b) Capsule
   (c) Synconus
   (d) Drupe
3. Which one is not a berry fruit?
   (a) Banana
   (b) Papaya
   (c) Sapota
   (d) Bael
4. Which one is climateric fruit?
   (a) Pineapple
   (b) Ber
   (c) Annona
   (d) Litchi
5. Which fruit shows duodichogamy?
   (a) Avocado (PDSD)
   (b) Chestnut
   (c) Pecanut
   (d) Cashew nut
6. Kiwi fruit belongs to which category?
   (a) Auto-octaploid
   (b) Allo-octaploid
   (c) Auto-hexaploid
   (d) Allo-hexaploid
7. Which citrus fruit is from the new world?
   (a) Grape Fruit
   (b) Pummelo
   (c) Lime
   (d) Sweet orange

Bold Means Correct Answer

8. Which fruit is called commonly "Bathroom Fruit"
   (a) Banana (b) Kiwi
   (c) Guava **(d) Mango**

9. Which fruit is called "Chinese Gooseberry"
   (a) Fig **(b) Kiwi**
   (c) Aonla (d) Strawberry

10. Which fruit is called "Indian Gooseberry"
    (a) Fig (b) Kiwi
    (c) Aonla (d) Strawberry

11. Which sub-tropical fruit was introduced in India under name of 'Japanese Medlar'?
    (a) Peach (b) Plum
    (c) Loquat (d) Quince

12. Propagation method in cashew nut
    (a) Softwood grafting (b) Hardwood cutting
    (c) Seedling (d) Patch budding

13. Inflorescence of arecanut is:
    (a) Spadix (b) Corymb
    (c) Solitary (d) Panicle

14. Edible part of banana is:
    (a) Mesocarp (b) Endocarp
    (c) Pericarp **(d) Both a & b**

15. Variety as a pollinizer used in mango is:
    (a) Chausa (b) Mallika
    (c) Bombay Green (d) Dasehari

16. Apple shape cultivar of mango is:
    (a) Rosica **(b) Rumani**
    (c) Totapuri (d) Tomy Atkin

17. Suitable temperature for banana cultivation is:
(a) 17-20°C **(b) 20-30°C**
(c) 30-34°C (d) 8-15°C

18. Amritsagar is important cultivar of:
(a) Mango (b) Bael
(c) Banana (d) Guava

19. Papaya is commercially propagated by:
(a) Softwood *cutting* **(b) Seeds**
(c) Patch budding (d) Grafting

20. Enzyme present in pineapple is:
(a) Papain (b) Caprine
(c) Bromelin (d) Bromacil

21. Adam's Apple is important rootstock of:
(a) Apple (b) Fig
(c) Sapota (d) Almond

22. Jackfruit is well acclimatized in which type of climate
(a) Temperate **(b) Tropical**
(c) Sub-tropical (d) Arid zone

23. Carambola is native of:
(a) Japan **(b) Indonesia**
(c) Thailand (d) China

24. Chemical used for control for pre- harvest drop of citrus fruits is:
(a) 2-4 D @ 10 ppm (b) 2-4 D @ 200 ppm
(c) NAA @ 50 ppm (d) NAA @ 500 ppm

25. Star Ruby is popular & commercial cultivar of:
(a) Pomegranate **(b) Grape fruit**
(c) Strawberry (d) Peach

26. Vines of grapes are planted in the northern plains of India in:

(a) Oct-Nov **(b) Dec-Jan**

(c) Jan-March (d) July-August

27. Grape cultivar resistant to anthracnose

(a) Pusa Navrang (b) Arka Neelmani

(c) Pusa Urvashi (d) Arka Majestic

28. Gulabi is important cultivar of:

(a) Grape Fruit (b) Pomegranate

(c) Strawberry **(d) Litchi**

29. Fruit cracking in pomegranate is more take place in:

(a) Ambe Bahar **(b) Mrig Bahar**

(c) Haste Bahar (d) All

30. Time of fruit bud differentiation in Anola is:

(a) Jan-Feb **(b) March-April**

(c) May-June (d) July-August

31. Date is which type of fruit?

(a) Drupe (b) Stone

(c) Berry (d) Capsule

32. Bael is richest source of:

(a) Vitamin-E (b) Vitamin-C

(c) Vitamin-B complex (d) Vitamin-A

33. Origin of *Carrisa carandas* is from:

(a) India (b) Africa

(c) Europe (d) Brail

34. Origin of *Carrisa grandiflora* is from:

(a) India **(b) Africa**

(c) Europe (d) Brazil

35. Which state is known as" "Apple Bowl?

(a) HP (b) J & K
(c) Uttarakhand (d) None

36. AKBAR is new cultivar of which fruit crop?

(a) Apple (b) Peach
(c) Plum (d) Apricot

37. Scientific name of pear:

(a) *Pyrus serotina* (b) *P. pashia*
(c) *P. Pyrifolia* **(d) *P. communis***

38. Flower colour of peach is:

(a) White (b) yellow
(c) Pink (d) Red

39. Dwarf rootstock of plum is:

(a) Paja (b) Colt
(c) Pixy (d) Myrobalan-B

40. Chilling requirement of almond is:

(a) 800 hrs (b) 1000 hrs
(c) 1500 hrs (d) 2700hrs

41. Which one is best cultivar of apricot for outdoor cultivation in kitchen garden?

(a) New Castle **(b) Moorpark**
(c) Nugget (d) Shakarpara

42. Largest producer of raw cashew nut in the world is:

(a) USA (b) Brazil
(c) India (d) China

43. Apple is divided in how much grades

(a) 6 (b) 3
(c) 4 (d) 26

44. Most common tree used for shade purpose in coffee is:

(a) Silver Oak (b) Popular

(c) Eucalyptus (d) All of above

45. Fruit of Areca nut is:

(a) Drupe **(b) Berry**

(c) Stone (d) Nut

46. Rubber tree is native of:

(a) Brazil (b) China

(c) Malaysia (d) India

47. Popular sweet drink obtains from the sugary palm (Palmyra Palm) is:

(a) Toddy (b) Fenny

(c) Neera (d) Perry

48. Major disease of grape is:

(a) Downy mildew (b) Powdery mildew

(c) Pink berry formation (d) Blossom drop

49. Which one is serious pest of apple?

(a) Wooly apple aphid (b) Sanjose scale

(c) Fruit fly (d) Shoot & fruit borer

50. Calyx end rot of grapes is cause due to deficiency of:

(a) **Ca** (b) B

(c) Zn (d) Mg

51. Little leaf of coconut is caused due to:

(a) MLO's **(b) Zn deficiency**

(c) Virus (d) Fungal

52. Walnut is harvested at which stage:

(a) PTB or full ripe stage (b) Half maturity

(c) Green colour stage (d) None

53. Highest chilling requirement among the temperate fruits:

(a) Plum (b) Peach
(c) Apricot **(d) Cherry**

54. Self unfruitful cultivar of peach is:

(a) J.H.Hale (b) Red Heaven
(c) Sharanpur Prabhat (d) Flordasum

55. Chromosome number of Coconut

(a) 42 (b) 40
(c) 36 **(d) 32**

56. Grit cell free variety of pear is:

(a) Magness (b) Flemish beauty
(c) Both a & b (d) Bartlett

57. Which variety is used as pollinizer for Red Delicious cultivar of apple?

(a) Starkllng Delicious (b) Golden Delicious
(c) Royal Delicious **(d) Red Gold**

58. Which one is ultra dwarf rootstock of apple?

(a) M-9 (b) M-13
(c) M- 27 (d) MM-111

59. Which fruit is rich in malic acid?

(a) Citrus (b) Guava
(c) Pineapple **(d) Cherry**

60. Custard apple is which kind of fruit botanically?

(a) Sorosis (b) Synconus
(c) Eteario of berries (d) Eteario of achenes

61. Which cultivar of mango has almost seedless fruit due to stenospermocarpy?

(a) Neelum **(b) Sindhu**
(c) Ratna (d) Alphonso

62. Which fruit is commonly known as "Finest fruit of world?

(a) Avocado **(b) Mangosteen**

(c) Kiwi (d) Persimmon

63. June drop in apple is due to:

(a) Improper pollination & fertilization

(b) Water deficiency

(c) High Temperature

(d) All of above

64. Black tip of mango is caused by:

(a) B deficiency (b) Zn deficiency

(c) K deficiency (d) Ca deficeincy

65. Pear decline is caused by:

(a) Fungus (b) Virus

(c) Bacteria **(d) MLO's**

66. Mango canker is caused by:

(a) Fungus (b) Virus

(c) Bacteria **(d) Algae**

67. Which one is highest edible oil yielding crop perennial crop?

(a) Oil Palm (b) Ground nut

(c) Coconut (d) Sunflower

68. Which one is shows cauliflorous bearing habit?

(a) Coffee **(b) Cocoa**

(c) Oil Palm (d) All the above

69. Rust resistance variety of coffee is:

(a) Cauvery (b) Kent

(c) Sanraman (d) None of above

70. Best harvesting or plucking stage of tea is:

(a) Two leaves, two bud stage

(b) Two leaves, single bud stage

(c) Single leaf, single bud stage

(d) All of above

71. Largest producer of coconut

(a) Indonesia (b) Philippines

(c) Malaysia (d) India

72. Which fruit has good amount of phosphoric acid?

(a) Almond (b) Carambola

(c) Walnut (d) Pecan nut

73. Apple cultivar free from all fungal diseases

(a) Freedom **(b) Liberty**

(c) Tropical Beauty (d) Ambri

74. Macadamia a minor fruit crop is introduction from:

(a) Australia (b) USA

(c) Japan (d) Brazil

75. Band disease of arecanut is due to:

(a) B deficiency (b) Ca deficiency

(c) Improper drainage (d) MLO's

76. little leaf of litchi is due to;

(a) Zn deficiency (b) MLO

(c) Fungus (d) None

77. Serious pest of litchi is:

(a) Aphid **(b) Mite**

(c) White fly (d) Bark eating caterpiller

78. Which palm is source of sweetening drink?

(a) Oil palm (b) Date palm

(c) Palmyra palm (d) Coconut

79. RRII is important cultivar of which crop?

(a) Oil palm **(b) Rubber**

(c) Cashew nut (d) Cocoa

80. Which one is interspecific variety of arecanut?

(a) Sri Vardhan (b) Mangla

(c) Sumangala (d) Mohit nagar

81. Coffee is rich source of which vitamin?

(a) Thiamine (b) Riboflavin

(c) Niacin (d) None

82. Plum is divided in how much grades?

(a) 3 (b) 4

(c) 6 (d) 26

83. Vivipary is common in which fruit crop?

(a) Grape fruit (b) Cocoa

(c) Jackfruit **(d) All of above**

84. Juice content present in citrus Is:

(a) 10-15 % (b) 20-30 %

(c) 35-50 % (d) 65-75 H

85. Benzoic acid is most effective against.

(a) Bacteria (b) Mould

(c) Yeast (d) All of above

86. Acidity of jam and jelly is:

(a) 0.3% **(b) 0.5-0.7 %**

(c) 1.0 % (d) 12 %

87. Papaya has vitamin-A (in IU/100 g) content

(a) 4800 (b) 4000

(c) 3300 **(d) 2020**

88. Which of following fruit crop is cross pollinated?

(a) Date palm (b) Papaya

(c) Kiwi **(d) All of above**

89. Gola cultivar of ber has ploidy level.

(a) 2x (b) 4x

(c) 6x **(d) 8x**

90. Which fruit is commonly called "Ambassador of India"?

(a) Mango (b) Banana

(c) Apple (d) Guava

91. Melon tree is common name of which fruit?

(a) Peach (b) Plum

(c) Loquat **(d) Papaya**

92. Best method for top working in Mango is:

(a) Veneer grafting **(b) Side grafting**

(c) Epicotyle grafting (d) Softwood grafting

93. Inflorescence of fig is:

(a) Spadix **(b) Hypanthodium**

(c) Panicle (d) Cymose

94. Edible part of anola is:

(a) Mesocarp (b) Endocarp

(c) Epicarp **(d) Both a & b**

95. Specific gravity of fruits is measured by:

(a) Podometer **(b) Pycnometer**

(c) Tendrometer (d) None

96. Apple colour is cultivar of which fruit?

(a) Mango (b) Banana

(c) Guava (d) Apple

97. Removal of male bud after completion of female phase in banana is known as:

(a) Denavelling (b) Mattocking

(c) Ratooning (d) Desuckering

98. Amritsagar is important cultivar of:

(a) Banana (b) Mango

(c) Bael (d) Guava

99. Mother of all coloured cultivar of mango is:

(a) Mallika (b) Rumani

(c) Mulgoa (d) Madhulica

100. Seed rate of dioecious papaya for one hectare is:

(a) 150-200 g (b) 250-450 g

(c) 450-500 g (d) 1000-1200g

101. Most commonly used chemical for flowering of Pineapple is:

(a) Ethrel or CEPA (b) NAA or Planofix

(c) 2, 4-D (d) IBA

102. Mahua is rootstock of:

(a) Fig **(b) Sapota**

(c) Peach (d) Plum

103. Which fruit is viviparous in nature?

(a) Grape fruit (b) Jackfruit

(c) Cocoa **(d) All of above**

104. Which fruit is commonly known as "five corner fruit"?

(a) Persimmon (b) Durian

(c) Passion fruit **(d) Carambola**

105. Early season variety of sweet orange:

(a) Pineapple (b) Valencia

(c) Hamlin (d) Jaffa

106. Star Ruby, important cultivar of grape fruit is developed through mutation of:

(a) Duncan **(b) Hudson**

(c) Foster (d) Red Blush

107. Most widely cultivated cultivar of grape in India Is:

(a) Thomson Seedless (b) Parlette

(c) Pusa Seedless (d) Tas-e-Ganesh

108. Propagation in coconut is by:

(a) Sucker (b) Runner

(c) Seed (d) Offsets

109. Bread fruit has which kind of parthenocarpy:

(a) Vegetative **(c) Stenospermocarpy**

(c) Stimulative (d) None

110. Pollination in fig is by:

(a) Wasp (b) Honey bees

(c) House fly (d) Fig Weevil

111. Edible part of date palm is:

(a) Mesocarp **(b) Pericarp**

(c) Epicarp (d) None

112. Most stable sex expression in papaya

(a) Maleness **(b) Femaleness**

(c) Hermaphrodite (d) None

113. Pusa Arunima Important cultivar of mango is cross of:

(a) Alphonso x Sensational (b) Sensational x Amarpali

(c) Amarpalli x Sensation (d) Amarpalli x Bangalpalli

114. Leading commercial cultivar of banana is:

(a) Dwarf Cavendish (b) Poovan

(c) Nendran **(d) Grand Naine**

115. Papaya cultivar suitable for pot gardening is:

(a) Pusa Delicious **(b) Pusa Nanha**

(c) Pusa Majestic (d) Taiwan

116. Pusa Srijan is new rootstock of:

(a) Mango (b) Citrus

(c) Guava (d) None

117. Pusa Srijan important rootstock of guava is genetically:

(a) Diploid (b) Monosomic

(c) Trisomic **(d) Aneuploid**

118. Which fruit has anti-scurvy (vitamin-C deficiency disease) property?

(a) Bael (b) Jamun

(c) Karonda (d) Mango

119. Active ingredient present in Bael is:

(a) Ficin **(b) Marmelosin**

(c) Carpine (d) Aegleosin

120. In India date is harvested at which stage?

(a) Dang **(b) Doka**

(c) Pind (d) None

121. Vitamin-C present in aonla per 100 g:

(a) 40-90 mg (b) 150 mg

(c) 300 mg **(d) 600 mg**

122. Amlidana new cultivar of pomegranate is released from which research station?

(a) IIHR (b) IARI

(c) PDKV Akola (d) Pune

123. Variety of litchi mainly used for canning purpose is:

(a) Kasaba (b) Rose Scented

(c) Swarna Roopa **(d) Shahi**

124. Late maturing cultivar of grape is:

(a) Arka Shyam (b) Arka Hans

(d) Arka Kanchan (d) Arka Chitrah

125. Red Blush is important cultivar of:

(a) Pomegranate **(b) Grape fruit**

(c) Strawberry (d) Peach

126. Pineapple is famous cultivar of:

(a) Sweet orange (b) Pineapple

(c) Mandarin (d) Grape fruit

127. Rootstock for HDP in citrus is:

(a) Rangpur lime (b) Trifoliate orange

(c) Troyer citrange (d) Flying dragon

128. Which is indicator for Citrus canker:

(a) Acid lime **(b) Kagzi lime**

(c) Tahti lime (d) Sweet lime

129. Pruning intensity of cultivar Blue Bokhri of grapes is:

(a) 3-4 buds (b) 6-7 buds

(c) 8-10 buds (d) 10-14 buds

130. Which fruit crop has mycorhizal root network?

(a) Pomegranate **(b) Litchi**

(c) Mango (d) Banana

131. First processed fruit in the human civilization:

(a) Fig (b) Grape

(c) Apple (d) Apricot

132. Which fruit is commonly known as "Fruit of New World"?

(a) Mangosteen (b) Jackfruit

(c) Avocado (d) Passion Fruit

133. Which cultivar of date palm is developed by somatic embryogenesis?

(a) Khad awy (b) Medjool

(c) Halawy **(d) Both a & b**

134. What is scientific name of "Rose Apple"?

(a) *Syzygium jambos* (b) *S. unciflora*

(c) *S. cumini* (d) *S. densiflora*

135. Which one is typical sub-tropical evergreen pome fruit?

(a) Peach (b) Plum

(c) Apricot **(d) Loquat**

136. Tenturier (red skin and red flesh) cultivar of grape is:

(a) Arka Navrang (b) Arka Neelmani

(c) Arka Majestic (d) Arka Urvashi

137. Purbi is important cultivar of:

(a) Drupe Fruit (b) Pomegranate

(c) Strawberry **(d) Litchi**

138. Amlidana new cultivar of pomegranate is cross of:

(a) Ganesh x Nanha (b) Ganesh x Alandi

(c) Alandi x Jyoti (d) Ganesh x Gule Shah Red

139. Fruit bearing in aonla is start after how much time of planting?

(a) 1 year (b) 2 year

(c) 3 year (d) 4 year

140. A fine date palm variety requires how much units of heat for full maturity?

(a) 1600 (b) 2400

(c) 3300 (d) 4100

141. Which one is called "Holy Fruit"?

(a) Banana (b) Mango

(c) Bael (d) Jamun

142. Origin of *Carrisa grandiflora* (netal plum) is:

(a) India **(b) S. Africa**

(c) Europe (d) Brazil

143. Chilling hours requirement of apple is:

(a) 450 hrs (b) 800 hrs

(c) 1200 hrs (d) 1000 hrs

144. AKBAR new cultivar of apple is cross of:

(a) Ambri x Cox Orange Pippin

(b) Ambri x Baldwin

(c) Ambri x Pippin

(d) Ambri x Red Delicious

145. Important rootstock of pear Kainth is botanically knows as:

(a) *Pyrus serotina* **(b) *P. pashia***

(c) *P. Pyrifolia* (d) *P. communis*

146. Which temperate fruit require highest pruning for regular fruiting?

(a) Apricot (b) Pear

(c) Peach (d) Plum

147. TSS of plum is:

(a) 10° brix **(b) 12.5° brix**

(c) 16^0 brix (d) None

148. Natural hybrid Sloh is cross of:

(a) Peach x Almond (b) Almond x Peach

(c) Almond x Plum (d) Plum x Almond

149. Origin of apricot is from:

(a) South China **(b) East China**

(c) Japan (d) Europe

150. Cashew is cultivated at altitude of for proper fruiting:

(a) Less than 400 m **(b) Less than 700 m**

(c) Less than 1000 m (d) Less than 1200 m

151. Cashew is divided in how much grades for export purpose?

(a) 3 (b) 4

(c) 6 **(d) 26**

152. Chromosome no. of *Coffea arabica* is:

(a) 22 (b) 24

(c) 36 **(d) 44**

153. Processed green nuts of arecanut are called:

(a) Kalipak (b) Kottapak

(c) Chali (d) None

154. Propagation method In rubber is:

(a) Softwood grafting **(b) Forket budding**

(c) Patch budding (d) Chip budding

155. Alcoholic drink prepared from palmyra palm is-

(a) Toddy (b) Fenny

(c) Perry (d) Cidar

156. Downy mildew of grapes is caused by:

(a) Fungus (b) Bacteria

(c) Virus (d) MLO's

157. Notorious pest of apple is:

(a) Wooly apple aphid **(b) San jose scale**

(c) Fruit fly (d) Shoot & fruit borer

158. Coloure & Shot Berries formation in grapes is due to:

(a) Ca deficiency (b) Zn deficiency

(c) B deficiency (d) None

159. Band disease is physiological disorder of:

(a) Oil palm **(b) Arecanut**

(c) Coffee (d) Cocoa

160. *Pouteria compechiana* (egg fruit) belongs to which family?

(a) Spindaceae **(b) Sapotaceae**

(c) Bombaceae (d) Saxifereaceae

161. Chromosome no. of mango is:

(a) 40 (b) 42

(c) 38 (d) 48

162. One should consume how much leafy vegetable per day?

(a) 125g (b) 145g

(c) 200g (d) 250g

163. Which one of the following is not a pome fruit?

(a) Apple **(b) Peach**

(c) Pear (d) Loquat

164. Which of fruit has highest rate of ethylene production?

(a) Apple (b) Passion fruit

(c) Pear (d) Banana

165. Atemoya is cross of which annona species?

(a) *A. squamosa x A. cherimoya*

(b) *A. cherimoya x A. squamosa*

(c) *A. muricata x A. cherimoya*

(d) None

166. Cashew nut is native of:

(a) India (b) Africa

(c) USA **(d) Brazil**

167. Propagation in fig is by:

(a) Hardwood cutting (b) Softwood cutting

(c) T-budding (d) None

168. Pineapple has which type of parthenocarpy?

(a) Vegetative (b) Stimulative

(c) Stenospermocarpy (d) None

169. Banana has which type of parthenocarpy?

(a) Vegetative (b) Stimulative

(c) Stenospermocarpy (d) None

170. Pollination in mango is done by:

(a) Common Housefly (b) Honey bees

(c) House flies (d) Wind

171. Edible part of mango is:

(a) Placenta **(b) Mesocarp**

(c) Endocarp (d) Both a & b

172. Only European country which grows mango commercially?

(a) France **(b) Spain**

(c) Italy (d) Germany

173. Sai Sugandha is cross of:

(a) Totapuri x Kesar (b) Kesar x Totapuri

(c) Bombay x Kalapady (d) None

174. Seedlessness in banana is regulated by:

(a) NAA@25 ppm (b) Etylene@25 ppm

(c) 2. 4-D @25 ppm (d) IBA@25 pp

175. Papaya cultivar having pear shape fruits is:
   (a) Solo (b) Taiwan
   (c) Hatras Gold (d) Surya

176. Common rootstock of custard apple is:
   (a) *A. atemoya* **(b) *A. glabra***
   (c) *A. cherimoya* (d) *A. reticulata*

177. National fruit of Japan is:
   (a) Carambola **(b) Persimmon**
   (c) Mangosteen (d) Passion fruit

178. Which one is monoembryonic citrus?
   (a) Acid lime (b) Grape fruit
   (c) Lemon **(d) Pummelo**

179. Gajnimma is citrus rootstock having scientific name:
   (a) *C. pennivesiculata* (b) *C. assamensis*
   (c) *C. latifolia* (d) *C. liletoides*

180. Papain enzyme extracted from papaya has how much protein?
   (a) 22 % (b) 44 %
   (c) 60 % **(d) 72.2 %**

181. Typical example of parthenocarpic fruit development is:
   (a) Persimmon (b) Banana
   (c) Mangosteen (d) Grape

182. Jonathan spot of apple is due to deficiency of:
   (a) B (b) Ca
   (c) Zn **(d) Water**

183. Which cultivar of mango is prone to mango malformation?
   (a) Bombay Green (b) Bahaduran
   (c) Illaichi (d) Alphonso

184. Root wilt of coconut is caused by:

(a) Fungus (b) Bacteria

(c) Virus **(d) MLO's**

185. Mango rust is caused by:

(a) Virus (b) Fungus

(c) Bacteria **(d) Algae**

186. In oil palm removal of flowers in the early years of plantation is known as:

(a) Ablation (b) Denavelling

(c) Stripping (d) Skuffing

187. Horizontal branches of cocoa plants is known as:

(a) Jarquette (b) Chupan

(c) Plagiotropic (d) Orthotropic

188. Bearing in coffee starts after how much time of planting?

(a) 1 year (b) 2-3 years

(c) 3-4 years (d) 4-6 years

189. Best nitrogen fertilizer for tea crop:

(a) Urea **(b) CAN**

(c) Ammonium Nitrate (d) DAP

190. Amrita is new cultivar of which fruit crop released by KAU, Kerala?

(a) Pineapple (b) Coconut

(c) Aonla (d) Banana

191. Which fruit shows heterodichogamy?

(a) Chestnut (b) Walnut

(c) Pecannut (d) Cashewnut

192. Paradox is famous cultivar of:

(a) Apricot **(b) Walnut**

(c) Plum (d) None

193. Chilling requirement of cherry is:

(a) 400-700 hrs (b) 1000-1200 hrs

(c) 1500-1800 hrs **(d) 2000-2700hrs**

194. Sharanpur Prabaht important cultivar of peach is cross of:

(a) Sharbati x J.H.Hale

(b) Sharbati x Shane-e-punjab

(c) Sharbati x Flordasum

(d) Sharbati x Redheaven

195. Interstock used between Bartlett & Quince cultivar of pear is:

(a) Old Home (b) Beaurre

(c) Both a & b (d) Leconte

196. Commercial cultivar of apple in USA:

(a) Starkling Delicious **(b) Golden Delicious**

(c) Royal Delicious (d) Red Gold

197. M-27 ultra dwarf rootstock of apple is cross of

(a) M-13 x M-9 (b) M-13 x M-7

(c) MM-104 x M-9 (d) MM 104 x M- 7

198. Which Sub- tropical fruit introduced in India under the name of 'Japanese Medlar

(a) Peach (b) Plum

(c) Apricot **(d) Loquat**

199. Which rootstock of jamun is resistant to termites?

(a) *Syzygium jambos* (b) *S. unciflora*

(c) *S. densiflora* (d) None

200. Chuhhara making variety of date palm is:

(a) Khadrawy (b) Medjool

(c) Halawy **(d) Both a & b**

201. Fat percentage of avocado is:
(a) 20 % **(b) 24 %**
(c) 28 % (d) 42 %

202. Active compound present in fig is:
(a) Prunacin (b) Papain
(c) Ficin (d) Carpine

203. Fruit of litchi is botanically is:
(a) Drupe (b) Berry
(c) Capsule **(d) Nut**

204. Cap like structure calyptra is formed in grape by union of:
(a) Petals (b) Sepals
(c) Both a & b (d) Stamens

205. Pomology is derived from which language?
(a) Latin **(b) Greek**
(c) English (d) French

206. Chromosome no. of Papaya is:
(a) 16 **(b) 18**
(c) 24 (d) 32

207. Which of the following fruit belong to non-climateric fruits?
(a) Apple (b) Papaya
(c) Guava **(d) Cashewnut**

208. Most common method of cashew roasting
(a) Drum roasting **(b) Steam roasting**
(c) Oil bath roasting (d) None

209. Dudsagar is famous cultivar of:
(a) Mango **(b) Banana**
(c) Bael (d) Guava

210. Propagation in coffee is done by:

(a) Layering (b) Cutting

(c) Seed (d) Grafting

211. Which fruit has stimulative parthenocarpy?

(a) Banana (b) Grape

(c) Litchi (d) Fig

212. Permanent sod culture is common practice in orchard of:

(a) Pear (b) Peach

(c) Plum (d) Apricot

213. *Prunus behmi* is natural cross hybrid

(a) Almond x Peach (b) Peach x Almond

(c) Pear x Peach (d) Peach x Plum

214. Scientific name of Shaira, Important rootstock of pear:

(a) *Pyrus pashia* **(b) *P. serotina***

(c) *P. pyrifolia* (d) *P. communis*

215. In which fruit seed germinate within the fruits?

(a) Grape fruit (b) Jackfruit

(c) Both a & b (d) Jamun

216. Khirni is popular rootstock of:

(a) Sapota (b) Fig

(c) Almond (d) Walnut

217. TSS of pineapple is about:

(a) 8-10 % **(b) 10-12 %**

(c) 16-18 % (d) 20-22 %

218. Yellow pigment present in papaya is due to:

(a) Xanthophyll **(b) Caricaxanthin**

(c) Carotenoids (d) Flavonoids

219. Genetic constitution of Lady Finger cultivar of banana is:

(a) AAA (b) AAB

(c) AA **(d) AB**

220. Which is monocarpic and monocotyledonus fruit crop:

(a) Banana (b) Pineapple

(c) Both a & b (d) Strawberry

221. Mallika very famous cultivar of mango is cross of

(a) Neelum x Deshehari (b) Deshehari x Neelum

(c) Ratna x Alphonso (d) Neelum x Alphonso

222. VHT in mango is recommended for disinfection against which pest?

(a) Fruit fly (b) Stone weevil

(c) Mango hopper **(d) Both a & b**

223. Edible part of litchi is

(a) Fleshy stalk (b) Fleshy peduncle

(c) Fleshy arils (d) Fleshy thalamus

224. Inflorescence of mango is:

(a) Fascicle **(b) Panicle**

(c) Solitary (d) Spadix

225. Mchhntosh is which season cultivar of apple?

(a) Early **(b) Mid**

(c) Late (d) Off-season

226. Chilling requirement of apple stratification is:

(a) 0^0C (b) 4^0C

(c) 7.2^0C (d) 9^0C

227. *In-situ* method of propagation of mango is:

(a) Veneer grafting

(b) Softwood grafting (Start AAU, Guj.)

(c) Side grafting

(d) Inarching

228. Storage temperature of banana is:

(a) 8-9°C (b) 10-11°C

(c) 13°C (d) None

229. Which fruit is commonly known as "Adam's Fig"

(a) Banana (b) Mango

(c) Fig (d) Sapota

230. Which fruit is commonly known as "Adam's Apple"

(a) Banana (b) Mango

(c) Fig **(d) Sapota**

231. Polyploidy level of anola is:

(a) Diploid **(b) Autotetraploid**

(c) Amphidiploid (d) Allotetraploid

232. Which one is temperate fruit?

(a) Mango (b) Banana

(c) Cashewnut **(d) Walnut**

233. Aonla is botanically which kind of fruit?

(a) Nut (b) Stone

(c) Capsule (d) Berry

234. Origin of persimmon is from which country?

(a) China (b) Malaysia

(c) Japan (d) Taiwan

235. Which fruit is called "Apple of Tropics"?

(a) Ber **(b) Guava**

(c) Pomegranate (d) Sapota

236. Origin of litchi is from:

(a) India (b) East China

(c) South China (d) Brazil

237. Self sterile variety of almond is:

(a) Sloe (b) Drake

(c) Katha (d) Debhar

238. Storage temperature for mango is:

(a) 3-4°C (b) 5-6°C

(c) 8-9 °C (d) 13 °C

239. Cashew apple is botanically :

(a) Fruit **(b) Fleshy peduncle**

(c) Cotyledons (d) Fleshy thalamus

240. Hybridization in mango is first started by

(a) Woodhouse

(b) Burns & Prayag (1911, at Pune)

(c) Tanaka & Swingle

(d) Mukeerji

241. Amarpalli important rootstock of mango is cross of

(a) Deshehari x Neelum (b) Neelum x Deshehari

(c) Alphonso x Neelum (d) Neelum x Alphonso

242. Weight of sword sucker in banana is:

(a) 250 g (b) 450 g

(c) 750 g (d) 800-900 g

243. Enzyme present in papaya use in tenderization of meat:

(a) Papain (b) Trypsin

(c) Carpine (d) Carixanthin

244. No. of pineapple plants under HDP:

(a) 25.000 (b) 45.000

(c) 53,000 **(d) 64,000**

245. Which fruit show cauliflorus bearing habit?

(a) Papaya **(b) Jackfruit**

(c) Mango (d) Grape fruit

246. Acid percentage in sour type carambola is:

(a) 0.4% **(b) 1.0%**

(c) 1.2% (d) 1.8%

247. Blood Red is important cultivar of:

(a) Sweet orange (b) Grape fruit

(c) Pomegranate (d) Guava

248. Thompson is popular cultivar of:

(a) Grape (b) Strawberry

(c) Grapefruit (d) Coconut

249. Pruning in grape is done how many times in a year in Southern India?

(a) Once **(b) Twice**

(c) Thrice (d) No Pruning

250. Generally TSS of grape is:

(a) 12-14 % (b) 14-16 %

(c) 20-24 % (d) 24-28 %

251. Rose Scented is popular cultivar of:

(a) Apple (b) Peach

(c) Pomegranate **(d) Litchi**

252. Dwarf cultivar of pomegranate suitable for HDP is:

(a) Nanha (b) Ganesh

(c) Mridula (d) Jyoti

253. Highest vitamin-C is present in which fruit?

(a) Aonla (b) Litchi

(c) Barbados cherry (d) Guava

254. Resinous canal is main feature of which family?

(a) **Anacardiaceae** (b) Rosaceae

(c) Rutaceae (d) Myrtaceae

255. Tall mutant of Dwarf Cavendish is:

(a) **Grand Naine** (b) Robusta

(c) Nendran (d) Lady Fingar

256. Hatras Gold is important cultivar of:

(a) Banana (b) **Papaya**

(c) Guava (d) Mango

257. Guava wilt is more common in which kind of soils?

(a) Acidic (b) **Alkali**

(c) Alkaline (d) All of above

258. *Annona atemoya* is synthetic hybrid is cross of:

a) *Annona squamosa* x *Annona muricata*

b) *Annona squamosa* x *Annona reticulata*

c) *Annona squamosa* x *Annona cherimoya*

d) *Annona squamosa* x *Annona glabra*

259. Astringent variety of persimmon is:

(a) Fuyu (b) Jiro

(c) 20th Centaury (d) **Hachiya**

260. Kinnow hybrid is developed by:

(a) **H.B. Frost** (b) H.D. Olmo

(c) J. H. Higgin (d) Raja Bhosle

261. Who is the grape breeder:

(a) H.B. Frost (b) **H.P. Olmo**

(c) J. H. Higgin (d) Raja Bhosle

262. Adjamir is popular rootstock of:

(a) Guava **(b) Citrus**

(c) Apple (d) Almond

263. Which one is polyembryonic citrus?

(a) Pummelo **(b) Acid lime**

(c) Tahiti lime (d) Citron

264. Pollination in papaya is by:

(a) Wind (b) Human

(c) Insect **(d) Both a & b**

265. Which one is monoembryonic citrus?

(a) Pummelo (b) Tahiti lime

(c) Citron **(d) All the above**

266. Edible part of ber is:

(a) Pericarp (b) Mesocarp

(c) Endocarp (d) Epicarp

267. Little leaf of mango is caused due to:

(a) Leaf hopper (b) B deficiency

(c) Zn deficiency (d) MLO

268. Pusa Surya is new cultivar of:

(a) Mango (b) Papaya

(c) Guava (d) Aonla

269. Surya is a cultivar of:

(a) Mango **(b) Papaya**

(c) Guava (d) Aonla

270. Gandevi banana is selection from:

(a) Rasthali **(b) Basrai**

(c) Bombay Green (d) Laden

271. Dioecious variety of papaya is:

(a) Pusa Majesty (b) **Pusa Nanha**

(c) Pusa Delicious (d) Sunrise Solo

272. Aneuploid- 82 (Pusa Srijan) is dwarfing rootstock of:

(a) Guava (b) Papaya

(c) Mango (d) Orange

273. Which annona is commonly known as sweet Sop?

a) *A. squamosa* (b) *A. reticulata*

(c) *A. atemoya* (d) *A. muricata*

274. Persimmon has gained reputation as national fruit of Japan but is native to which country?

(a) India (b) Malaysia

(c) China (d) Indonesia

275. Physiological disorder of strawberry is:

(a) Albinism (b) Multiple crown

(c) Flattening of leaves (d) Fruit cracking

276. Highest mean bunch weight of grape is obtained under which training system of grapes?

(a) Head (b) Kniffin

(c) Bower (d) Telephone

277. Fig belongs to which family?

(a) Moraceae (b) Apocynaceae

(c) Rosaceae (d) Caricaceae

278. Sugars present in the Avocado is:

(a) 0.1 % **(b) 1.0 %**

(c) 10 % (d) 20 %

279. Best time of pruning of date palm in the north-plains is:

(a) Jan (b) March

(c) May (d) July

280. *S. densiflora* popular rootstock of Jamun is resistant against:

(a) Salinity (b) Nematodes

(c) Termite (d) Drought

281. Fire Ball is important cultivar of:

(a) Peach (b) Plum

(c) Pear **(d) Loquat**

282. Drought resistant rootstock of apple is:

(a) MM-104 (b) MM-111

(c) Merton-793 (d) M-9

283. Dwarf rootstock of apple is:

(a) MM-111 (b) MM-101

(c) Merton-793 **(d) M-9**

284. Triploid cultivar of apple is:

(a) Ambri **(b) Baldwin**

(c) Richard (d) Crimson Gold

285. Punjab Nectar is famous cultivar of:

(a) Apple (b) Plum

(c) Pear (d) Peach

286. Peach start fruit bearing after:

(a) 1 year **(b) 2 years**

(c) 3 years (d) 4 years

287. Wine Kirschwascher is produced from pulp of which fruit?

(a) Cherry (b) Apple

(c) Pear (d) Grapes

288. Phytophthora & Nematode resistance rootstock of citrus is:

(a) Rangpur lime (b) Rough lemon

(c) Jatti-khati **(d) Trifoliate orange**

289. *J. hindasi* is important rootstock of:

(a) Apricot **(b) Walnut**

(c) Almond (d) Plum

290. Which state share highest production of coconuts?

(a) Kerala (b) Karnataka

(c) Andhra Pradesh (d) Maharashtra

291. What is the main objective of withering in processing of tea?

(a) Fermentation **(b) Removal of moisture**

(c) Cutting (d) None

292. Coffee reached at maturity after flowering of:

(a) 6 months **(b) 8 months**

(c) 10 months (d) 12-13 months

293. Oil palm having shell is known as:

(a) Dura (b) Pisifera

(c) Tenera (d) None

294. Major physiological disorder of cocoa is

(a) Little leaf (b) Yellow leaf disease

(c) Black pod (d) Die-back

295. Panama wilt of banana is mostly found in:

(a) Acidic soils (b) Alkaline soils

(c) Alkali soils (d) All soils

296. Which cultivar of mango is resistant to mango malformation?

(a) Bahaduran (b) Illachi

(c) Alib **(d) All of above**

297. Oil palm having shell is absent known as:

(a) Dura **(b) Pisifera**

(c) Tenera (d) None

298. Edible part of mangosteen is:
(a) Mesocarp (b) Pericarp
(c) Succulent thalamus **(d) Aril**

299. Coffee is native of which place?
(a) India (b) Brazil
(c) Abyssinia (d) Europe

300. Which fruit is known as king of temperate fruits?
(a) Apple (b) Walnut
(c) Peach (d) Almond

301. 20th Century is famous cultivar of:
(a) Coconut **(b) Persimmon**
(c) Passion Fruit (d) Carambola

302. Gajanimma (*C. pennivesiculata*) is very popular rootstock of:
(a) Citrus (b) Guava
(c) Peach (d) Plum

303. Common male parents use in grape breeding to transmit seedlessness:
(a) Queen of Vineyard (b) Pusa Seedless
(c) Sharad Seedless **(d) Both a & b**

304. Vitamin-C presents in litchi/100g:
(a) 150 mg (b) 300 mg
(c) 60 mg (d) 120 mg

305. Notching is popular practice performed in:
(a) Fig (b) Tea
(c) Banana (d) Grapes

306. Which fruit is recommended as high energy fruit for diabetic patient?
(a) Banana **(b) Avocado**
(c) Passion fruit (d) Jackfruit

307. Partial removal of bark above the bud is known as:

(a) Notching (b) Nicking
(c) Ringing (d) Dehorning

308. Partial removal of bark bellow the bud is known as:

(a) Notching **(b) Nicking**
(c) Ringing (d) Dehorning

309. Which fruit is extremely drought tolerant?

(a) Ber (b) Pomegranate
(c) Bael **(d) All of above**

310. Large size fruit Jamun cultivar is :

(a) Ra-jamun **(b) Paras (Guj)**
(c) Narendra (d) Desi jamun

311. California Advance & Tanaka is late season cultivar of:

(a) Pear (b) Peach
(c) Plum **(d) Loquat**

312. Most vigorous rootstock of apple is:

(a) M-27 **(b) Merton-793**
(c) MM-104 (d) MM-111

313. Indigenous cultivar of apple grown in Kashmir is:

(a) Rymer (b) Ambri
(c) Both a & b (d) Red Delicious

314. Punjab Gold and Punjab Nectar is famous cultivar of:

(a) Pear (b) Peach
(c) Plum (d) None

315. Open vase like training system is popular in:

(a) Pear **(b) Peach**
(c) Plum (d) None

316. Physiological disorder of mangosteen is:
(a) Apolexy **(b) Gamboge**
(c) Dry neck (d) Little leaf

317. Spongy tissue is main problem of which mango cultivar:
(a) Alphonso (b) Bombay Green
(c) Ratna (d) Langra

318. Leaf rust of coffee is caused by:
(a) Algae (b) Bacteria
(c) Fungus (d) Virus

319. Which cultivar of banana is susceptible to panama wilt?
(a) Gross Michel (b) Basrai Dwarf
(c) Poovan (d) None

320. Oil is extracted from which part of oil palm?
(a) Epicarp **(b) Mesocarp**
(c) Endocarp (d) None

321. Forestero is famous cultivar of:
(a) Coffee (b) Tea
(c) Cocoa (d) Oil Palm

322. Highest productivity of coconut is in which country?
(a) Indonesia (b) Malaysia
(c) Philippines **(d) India**

323. Clonal rootstock of cheery is:
(a) Pixy (b) Paja
(c) Colt (d) Mazzard

324. Seedling rootstock of cheery is:
(a) Pixy (b) Paja
(c) Mazzard **(d) All the above**

325. Passion fruit is rich source of:
   (a) **Vitamin A** (b) Vitamin B
   (c) Vitamin C (d) None

326. On the basis of photoperiodism papaya is:
   (a) SDP (b) LDP
   (c) **Neutral Plant** (d) None

327. Metaxenia effect is commonly associated which fruit crop?
   (a) Ber (b) **Date palm**
   (c) Coconut (d) Oil palm

328. Seven large is very famous cultivar of:
   (a) Coconut (b) Peach
   (c) **Bael** (d) Jackfruit

329. Percentage of pollinizer required in apple orchard for regular cropping
   (a) 5-10 % (b) **11-33 %**
   (c) 50 % (d) 67 %

330. Scab resistant variety of apple is:
   (a) Prima (b) Liberty
   (c) Freedom (d) **All of above**

331. Mehal is important rootstock of which fruit crop?
   (a) Peach (b) **Pear**
   (c) Apple (d) Plum

332. Prunacin principle glycoside present in pulp of:
   (a) Pear (b) Plum
   (c) **Peach** (d) Both a & b

333. Ne Plus Ultra is most important cultivar of:
   (a) **Almond** (b) Apricot
   (c) Peach (d) None

334. Which temperate fruit is highly perishable in nature?

(a) Cherry **(b) Apricot**

(c) Peach (d) Plum

335. Baby is grade category of which fruit?

(a) Pomegranate (b) Mango

(c) Pineapple (d) Apple

336. Scuffing is important cultural practice performed in:

(a) Oil palm **(b) Coffee**

(c) Tea (d) Cocoa

337. Indigenous cultivar of arecanut is:

(a) Sree Vardhani (b) Mangla

(c) Sumangla **(d) Mohitnagar**

338. Skiffing is important cultural practice performed in:

(a) Oil palm (b) Coffee

(c) Tea (d) Cocoa

339. Major physiological disorder of rubber is:

(a) Black pod **(b) Brown blast**

(c) Yellow disease (d) Root wilting

340. Alcohol percentage in toddy:

(a) 5 % (b) 10 %

(c) 14 % (d) 42 %

341. Moko disease of banana is caused by:

(a) MLO (b) Virus

(c) Bacteria (d) Fungus

342. Tea mosquito bug is major pest problem of:

(a) Tea **(b) Cashew nut**

(c) Coconut (d) Areca nut

343. Fruit necrosis of aonla is due to deficiency of:

(a) Zn (b) Ca

(c) B (d) None

344. Which part of *Bassia latifolia* is rich source of sugar:

(a) Corolla (b) Anthers

(c) Calyx (d) Fruit

345. Inflorescence of cashew nut is:

(a) Polygamomonecious (b) Fasicle

(c) Panicle (d) Spadix

346. New Castle is well known cultivar of:

(a) Peach (b) Plum

(c) Apricot (d) Loquat

347. Most commonly grown varieties of plum in India are:

(a) Japanese (b) European

(c) Hybrids (d) None

348. Which temperate fruit require lowest chilling requirement?

(a) Cherry **(b) Peach**

(c) Pear (d) Plum

349. Which clones are resistant to sigatoka leaf spot disease?

(a) AAA (b) AAB

(c) ABB (d) All of above

350. Coffee rust resistant species of coffee is:

(a) *C. arabica* (b) *C. robusta*

(c) *C. liberica* (d) *C. canephora*

351. Improper bunch filling in banana is due to deficiency of:

(a) B **(b) K**

(c) Zn (d) P

352. Apolexy is physiological disorder of:

(a) Pear (b) Persimmon

(c) Avocado **(d) Apricot**

353. Which one is tropical fruit crop:

(a) Custard apple (b) Citrus

(c) Walnut (d) Date Palm

354. Which one of fruit is not a stone fruit?

(a) Pear (b) Peach

(c) Plum (d) Cherry

355. Origin of sweet orange is:

(a) Brazil (b) India

(c) China (d) Europe

356. Persimmon is national fruit of:

(a) China **(b) Japan**

(c) Thailand (d) Indonesia

357. Sapota is belongs to which climate?

(a) Temperate **(b) Tropical**

(c) Sub-tropical (d) Arid zone

358. In fruit morphology carambola comes under which category?

(a) Capsule (b) Berry

(c) Nut (d) Drupe

359. Which fruit is native to India?

(a) Bael (b) Jackfruit

(c) Phalsa **(d) All of above**

360. Grape is rich source of:

(a) Folic acid **(b) Tartaric acid**

(c) Oxalic acid (d) Citric acid

361. Which fruit is commonly known as "Apple of Poor"?

(a) Guava (b) Banana
(c) Pear (d) Mango

362. Suitable storage temperature for apple is:

(a) 0°C (b) 0-2°C
(c) -2 to 0°C (c) -5°C

363. Inflorescence of ber is:

(a) Fasicle (b) Panicle
(c) Solitary (d) Spadix

364. Edible part of pineapple is:

(a) Mesocarp (b) Perianth
(c) Bracts **(d) Both b & c**

365. Caging technique of breeding in mango was first used by:

(a) Mukerjee (b) Brunus & Pryag
(c) Dr. R.N.Singh (d) Cheema & Dhani

366. Ratna important cultivar of mango is cross of:

(a) Neelum x Alphonso (b) Alphonso x Neelum
(c) Dashehari x Amarpalli (d) Amarpalli x Desahari

367. Major disease of HDP banana is:

(a) Sigatoka **(b) Finger tip**
(c) Panama wilt (d) Bunchy top

368. Genetic constitution of Gold Finger cultivar of banana is:

(a) AAAA (b) AABB
(c) AAAB (d) ABBB

369. HDP Suitable cultivar of papaya is:

(a) Pusa Delcious (b) Pusa Majesty
(c) CO-1 **(d) Pusa Nanha**

370. Multiple crown disorder of pineapple is common in which cultivar?

(a) **Kew** (b) Sugar Loaf
(c) Queen (d) Jaldhup

371. Sapota is propagated commercially by:

(a) Seed (b) **Inarching**
(c) Air layering (d) T-budding

372. Jack fruit is good source of:

(a) Protein (b) **Pectin**
(c) Fiber (d) Tannin

373. Marsh Seedless is seedless cultivar of:

(a) Sweet orange (b) Lemon
(c) Mandarin (d) **Grape fruit**

374. Back pruning or foundation pruning in grape is practiced in months of:

(a) **April** (b) July
(c) October (d) December

375. Juice content present in grapes is:

(a) 35 % (b) 45 %
(c) 55 % (d) **70 %**

376. Which mineral improve the quality of litchi fruit?

(a) B (b) **Ca**
(c) Zn (d) K

377. Spanish Ruby is important cultivar of:

(a) Grape fruit (b) Strawberry
(c) **Pomegranate** (d) None

378. Star Ruby is important cultivar of:

(a) **Grape fruit** (b) Strawberry
(c) Pomegranate (d) None

379. Propagation method of aonla is:

(a) T-budding
(b) Patch budding
(c) Ring budding
(d) Both a & b

380. A popular cultivar liquor prepared from date palm is:

(a) Arrack
(b) Dibbis
(c) Perry
(d) Cidar

381. Storage temperature of bael at RH 85-90%:

(a) 0°C
(b) 3°C
(c) 6-7°C
(d) 9°C

382. Most popular cultivar of kiwi fruit in the world is:

(a) Tomuri
(b) Monty
(c) Heyward
(d) Allison

383. June drop in apple is due to:

(a) Moisture stress
(b) Lack of pollination
(c) High wind speed
(d) Heavy fruit bearing

384. Low chilling required cultivar of apple is:

(a) Neomi
(b) Thimma
(c) Tropical Beauty
(d) All of above

385. Vigorous rootstock of pear is:

(a) Quince-A
(b) Quince-B
(c) Quince-C
(d) Kainth

386. Cashew nut is belongs to which family?

(a) Rutaceae
(b) Rosaceae
(c) Anacardiaceae
(d) Moraceae

387. Scientific name of chestnut is:

(a) *Castania sativa*
(b) *Corylus avellena*
(c) *Corylus colurana*
(d) None

388. Stone Fruit is physiological disorder of:

(a) Sapota | **(b) Custard apple**
(c) Avocado | (d) Persimmon

389. Aonla cultivar most prone to fruit necrosis:

(a) Francis | (b) Chakiya
(c) Both a & b | (d) Banarasi

390. White stem borer is major pest problem of:

(a) Oil palm | (b) Coconut
(c) Coffee | (d) Cashew nut

391. Pineapple wilt is caused by:

(a) Bacteria | **(b) Virus**
(c) Fungus | (d) Algae

392. Tapping panel dryness (TPD) is major physiological disorder of:

(a) Rubber | (b) Coconut
(c) Palmyra Palm | (d) Oil Palm

393. Sree Mangla is famous cultivar of:

(a) Coconut | (b) Coffee
(c) Arecanut | (d) None

394. Most widely grown coffee is:

(a) *Coffea arabica* | (b) *C. robusta*
(c) *C. canefera* | (d) *C. congenesis*

395. Stimulant present in tea is:

(a) Theol | **(b) Thein**
(c) Tannin | (d) Theaflavin

396. Most popular cultivar of kiwi throughout the world is:

(a) Tomuri | (b) Monty
(c) Allison | **(d) Heyward**

397. Sanjose Scale in apple got its entry from which country?

(a) Australia (b) USA

(c) France (d) England

398. Which temperate fruit comes earliest in flowering?

(a) Peach (b) Plum

(c) Cherry (d) Loquat

399. Which plum cultivar comes earliest in market?

(a) Jamuni (b) Santa Rosa

(c) Fronteir **(d) Titron**

400. Which temperate fruit comes first in market?

(a) Peach (b) Plum

(c) Cherry (d) Loquat

401. Kiwi is which type of fruit crop?

(a) Deciduous tree **(b) Deciduous vine**

(c) Evergreen tree (d) Evergreen vine

402. Best pollinizer variety for Non Pareil cultivar of almond is:

(a) Peerless **(b) Ne Plus Ultra**

(c) Sloe (d) Katha

403. Nugget is famous cultivar of:

(a) Peach (b) Plum

(c) Loquat **(d) Apricot**

404. Moisture content present in dry cashew nut is:

(a) 1 % **(b) 2-4 %**

(c) 6-8 % (d) 10-12 %

405. Aroma of tea is due to presence of:

(a) Thein **(b) Theol**

(c) Tannin (d) Theaflavins

406. Sex form of *Coffea arabica* is:

(a) 2x (b) 6x

(c) 4x (d) 8x

407. Samrudhi is well known cultivar of:

(a) Arecanut (b) Oil palm

(c) Coffee (d) Coconut

408. Process of latex extraction from the rubber is known as:

(a) Tipping **(b) Tapping**

(c) Skuffing (d) Ablation

409. Which palm is known as sugary or toddy palm?

(a) Oil palm (b) Coconut

(c) Palmyra palm (d) None

410. Pierce disease of grapes is caused by:

(a) Nematode (b) MLO

(c) Virus **(d) Fungus**

411. Pineapple wilt is transmitted by:

(a) Mite (b) Moth

(c) Mealy bug (d) Aphid

412. Anab-e-Roga in arecanut is caused by:

(a) *Hamella vasatrix* **(b) *Gonoderma lucidum***

(c) *Tephrina deformens* (d) *Ravenellia emblica*

413. Fruit cracking in pomegranate is more in which bahar?

(a) Ambe bahar (b) Haste bahar

(c) Mrig bahar (d) All of above

414. Pistachio nut is belongs to which family?

(a) Anacardiaceae (b) Juglandiaceae

(c) Caricaceae (d) None

415. Which fruit crop has sporophytic self incompatibility?

(a) Mango (b) Aonla

(c) Cocoa **(d) All of above**

416. Number of plants of kinnow under HPD system;

(a) 400 (b) 1600

(c) 3000 (d) 6400

417. Origin centre of sapota is:

(a) Mexico (b) Brazil

(c) India (d) Iran

418. Small holders irrigated plant crop is:

(a) Date palm **(b) Oil palm**

(c) Palmyra palm (d) Coconut palm

419. Propagation method of tea plantation is by:

(a) Softwood cutting (b) Leaf Nod Cutting

(c) Forklet budding **(d) Both a & b**

420. Fruit crop which shows polyembryony?

(a) Mango (b) Jamun

(c) Citrus **(d) All of above**

421. Pollination in apple is mainly govern by:

(a) Wind **(b) Honey bees**

(c) House flies (d) Artificial pollination

422. Fruit bearing habit of mango is:

(a) Terminal (b) Auxillay

(c) Mix bud (d) Cauliflorus

423. Export variety of mango is:

(a) Bangainpalli (b) Neelum

(c) Dashehari **(d) Alphonso**

424. Most serious pest of mango is:

(a) Mango hopper (b) Fruit fly

(c) Mango mealy bug (d) Stone weevil

425. Which cultivar of banana is also popular as French plantain:

(a) Poovan (b) Ney Poovan

(c) Nendran (d) Pey Kunnan

426. Sex ratio of Pusa Delicious is:

(a) 1:10 (b) 1:20

(c) 1:50 **(d) 50:50**

427. Lucknow-49 (Sardar) is developed from Allahabad Safeda through:

(a) Chance selection (b) Mutation

(c) Hybridization (d) Mass selection

428. Scientific name of pond apple is:

(a) *A. squamosa* **(b) *A. glabra***

(c) *A. diversifolia* (d) *A. reticulata*

429. Which citrus fruit is resistant to greening:

(a) Sweet lime (b) Kagzi lime

(c) Lemon (d) Acid lime

430. Parent of Sharad Seedless cultivar of grape is:

(a) Kishmish Charni (b) Anab-e-Shahi

(c) Thomson seedless (d) Parlette

431. To promote rooting of air layering in litchi IBA is used:

(a) 20-100 ppm (b) 200-1000 ppm

(c) 2000-10000 ppm (d) None

432. Dwarf cultivar of sapota is:

(a) Kirti Bharti (b) Cricket Ball

(c) Pala (d) Kalipatti

433. Mridula is famous cultivar of:

(a) Guava **(b) Pomegranate**

(c) Grape (d) Grape fruit

434. Aonla is which kind of fruit tree under north Indian conditions?

(a) Evergreen (b) Semi-deciduous

(c) Deciduous (d) None

435. Softening of dates is taken place at which stage?

(a) Doka (b) Pind

(c) Dang (d) Gandora

436. Banana takes how many days for maturity?

(a) 70-90 days **(b) 90-120 days**

(c) 120-150 days (d) 150-180 days

437. Gulabi is popular is important cultivar of:

(a) Jackfruit (b) Grape

(c) Grape fruit (d) None

438. Which variety of sweet orange has rudimentary fruit?

(a) Mosami (b) Kinnow

(c) Washington Navel (d) Mudkhed

439. Which one is cane prune cultivar of grape?

(a) Bangalore Blue **(b) Thomson Seedless**

(c) Perlette (d) None

440. Which one is soft seeded variety of pomegranate?

(a) Paper Shell (b) Jyoti

(c) Ganesh **(d) All of above**

441. Which one is deciduous fruit tree under north Indian conditions?

(a) Aonla (b) Bael

(c) Ber **(d) All of above**

442. Moisture percentage in fresh date:

(a) 10-20 % (b) 30-40 %
(c) 50-60 % **(d) 70-80 %**

443. Kiwi cultivar(s) commonly used as male plants or pollinizer in kiwi orchards:

(a) Allison (b) Tomuri
(c) Both a & b (d) Bruno

444. Optimum temperature for pollen germination & fruit setting in apple is:

(a) 0°C (b) 4.2-7°C
(c) 15-18°C **(d) 21.1-26.7°C**

445. Chilling requirement of pear is:

(a) 800 hrs (b) 1000 hrs
(c) 1200 hrs (d) 1600 hrs

446. What is enology/oenology?

(a) Science of wine making
(b) Science of raisin making
(c) Science of brewing and fragmentation
(d) Science of drug making

447. Fruit rich source of vitamin-A is

(a) Mango (b) Papaya
(c) Bael (d) None

448. Golden Star is important variety of:

(a) Persimmon (b) Grapes
(c) Carambola (d) Peach

449. Nitrogen present in CAN is:

(a) 16 % (b) 20 %
(c) 25 % (d) 32 %

450. Citrus fruit having mallic acid is:

(a) Sweet orange **(b) Sweet lime**

(c) Lemon (d) Kagzi lime

451. Which fruit shows vegetative parthenocarpy?

(a) Banana (b) Fig

(c) Litchi **(d) Both a & b**

452. Pollination in walnut is by:

(a) Wind (b) Insect

(c) Birds (d) Both a & b

453. Edible part of sapota is:

(a) Endocarp **(b) Mesocarp**

(c) Epicarp (d) Pericarp

454. Famous cultivar of mango for pickle use is:

(a) Bombay Green (b) Swarnrekha

(c) Ramkela (d) Totapuri

455. Which one is man made hybrid:

(a) Atemoya (b) Strawberry

(c) Lemon **(d) Both a & b**

456. Chromosome no. of persimmon is:

(a) 56 **(b) 90**

(c) 60 (d) 48

457. Sai suganda is famous cultivar of:

(a) Mango (b) Litchi

(c) Lemon (d) Peach

458. ELISA test in banana is used in screening of:

(a) Fungus **(b) Virus**

(c) Bunchy top (d) Aphids

459. Hermaphrodite variety of papaya is:
(a) Sunrise Solo (b) Surya
(c) Coorge Honey Dew **(d) a & c**

460. Parthenocarpic cultivar of guava is:
(a) Lucknow -49 (b) Allhabad Safeda
(c) Allahabad Round (d) Allahabad Surkha

461. Which one physiological disorder of citrus?
(a) Canker (b) Greening
(c) Gummosis **(d) Granulation**

462. Black bean is physiological disorder of:
(a) Coffee (b) Tea
(c) Cocoa (d) None

463. Which one of fruit is not a berry botanically?
(a) Apple (b) Papaya
(c) Grape (d) Banana

464. Which of the following fruit type is a nut?
(a) Coconut **(b) Litchi**
(c) Cherry (d) Cocoa

465. Which one is synthetic (man-made) hybrid fruit?
(a) Strawberry (b) Atemoya
(c) Wilking **(d) Above of all**

466. Fruit having double sigmoid fruit curve:
(a) Fig (b) Pear
(c) Kiwi (d) Mango

467. Which fruit is commonly known as "Star fruit"?
(a) Phalsa (b) Apple
(c) Mango **(d) Carambola**

468. Average nuts per coconut tree is:
(a) 11 (b) 22
(c) 33 **(d) 44**

469. Largest producer of banana is:
(a) China **(b) India**
(c) Sri Lanka (d) Thailand

470. Inflorescence of mulberry is:
(a) Catkin (b) Corymb
(c) Spathe (d) Spadix

471. Plum is originated from which place?
(a) China (b) Japan
(c) Europe (d) Asia-Minor

472. Fruit having single sigmoid curve:
(a) Apple (b) Fig
(c) Grape (d) Guava

473. Which self incompatibility is not present in fruit crops?
(a) Homomorphic **(b) Heteromorphic**
(c) Gametomorphic (d) Sporophytic

474. Ber is fruit crop belongs to which climate?
(a) Tropical (b) Temperate
(c) Sub-tropical (d) Humid

475. Which fruit is known as "Butter fruit"?
(a) Avocado (b) Coconut
(c) Cashew nut (d) Walnut

476. Average yield of grape is (t/ha)
(a) 8.0 **(b) 16.0**
(c) 20 (d) 25.0

477. Largest producer of coconut is:

(a) India (b) Malaysia

(c) Indonesia (d) Thailand

478. Inflorescence of pear is:

(a) Corymb (b) Catkin

(c) Spadix (d) Cymose

479. Parianth / bract is edible part of:

(a) Litchi (b) Bread fruit

(c) Fig **(d) Pineapple**

480. Which cultivar of mango has highest perfect flower:

(a) Dashehari **(b) Rumani**

(c) Bombay Green (d) Langra

481. Which cultivar of mango is suitable for HDP?

(a) Ratna (b) Sindhu

(c) Amrapali (d) Alphonso

482. CO-1 cultivar banana has same character to:

(a) Hill Banana (b) Dwarf Cavendish

(c) Nendran (d) Poovan

483. Most devastatibg disease of papaya nursery is:

(a) Wilt **(b) Damping off**

(c) Foot rot (d) None

484. Cultivar of pineapple having alcoholic flavour:

(a) Kew (b) Mauritius

(c) Jaldhup (d) Queen

485. Pala is popular cultivar of:

(a) Peach (b) Litchi

(c) Plum **(d) Sapota**

486. Which one is only triploid citrus fruit?

(a) Pummelo **(b) Tahiti lime**

(c) Kagi lime (d) Acid lime

487. Ultra dwarf rootstock of citrus is:

(a) Troyer citrange (b) Adjamir

(c) Trifoliate orange **(d) Flying dragon**

488. Moat widely accepted training system in grapes in India is

(a) Head system **(b) Bower system**

(C) Kniffin system (d) Both a & b

489. Which one is hard seeded variety of pomegranate?

(a) Khandari (b) Alandi

(c) Both a & b (d) Ganesh

490. Physiological disorder of aonla is:

(a) Fruit necrosis (b) Fruit rotting

(c) Stone fruit (d) Cracking

491. One kg fully ripe dates provide how much energy?

(a) 3300 (b) 3150

(c) 2700 (d) 2300

492. Kiwi fruit's vine was planted first time in:

(a) IARI, New Delhi

(b) Lal Bagh, Bangalore

(c) IIHR, Bangalore

(d) Botanical Garden. Kolkata

493. Wine produced from apple is known as:

(a) Perry (b) Neera

(c) Cidar (d) Fenny

494. Lal Ambri popular cultivar of apple is cross of:

(a) **Red Delicious x Ambri**

(b) Starking Delicious x Ambri

(c) Ambri x Red Delicious

(d) Ambri x Starking Delicious

495. Most commonly used rootstock for pear is:

(a) **Quince A** (b) Quince B

(c) Quince C (d) Kainth

496. Best time of pruning in peach is in:

(a) Early winters (b) Late winters

(c) **Mid winters** (d) Winter-spring

497. President is important cultivar of:

(a) Pear (b) **Plum**

(c) Peach (d) Walnut

498. Sharkarpara is important cultivar of:

(a) Pear (b) Plum

(c) Peach (d) **Apricot**

499. Harvesting of cashew nut is done in months of:

(a) **Feb-March** (b) June-July

(c) Sept-Oct (d) Dec-Jan

500. Tea plantation show good growth in the soils of:

(a) Neutral (b) Alkali

(c) **Acidic** (d) Alkaline

501. Cauvery important hybrid of coffee is cross of:

(a) Kent x HDT (b) **Cattura x HDT**

(c) Cioccie x HDT (d) None

502. Betel vine is trained on which training system?

(a) **Bareja** (b) T-bar

(c) Pergola (d) Kniffin system

503. Percentage of sugars in neera a sweet drink tapped from the palmyra palm:

(a) 5 % (b) **12%**

(c) 15% (d) 22%

504. Ring rust of aonla is caused by:

(a) Bacteria (b) Algae

(c) **Fungi** (d) Virus

505. Yellow leaf disease of arecanut is caused by:

(a) Bacteria (b) Algae

(c) Fungi (d) **MLO's**

506. Calyx end rot is physiological disorder of:

(a) **Persimmon** (b) Apple

(c) Custard apple (d) Sapota

507. Gooseberry is belongs to which family?

(a) Betulaceae (b) Fagaceae

(c) Bombaceae (d) **Saxiferaceae**

508. Kagzi lime is indicator of:

(a) Greening (b) **Tristeza**

(c) Canker (d) None

509. Sex form of mango is:

(a) **Allo-tetraploid** (b) Autotetraplold

(c) Auto-octaploid (d) Diploid

510. Propagation material of date palm is:

(a) Offset (b) Runner

(c) Sucker (d) **Both a & c**

511. Which fruit has nuceller polyembryony type of apomixis?

(a) Jamun (b) Citrus

(c) Both a & b (d) Mango

512. Pusa Delicious variety of papaya falls under which category?

(a) Monoecious **(b) Gynodioecious**

(c) Dioecious (c) Andromonocious

513. Off season cultivar of mango is:

(a) Madhulica **(b) Niranjan**

(c) Neelum (d) Rosica

514. Banana cultivar having unique aroma and flavour of apple:

(a) Virupakshi (b) Basrai

(c) Safed Valchi (d) Bombay Green

515. Lalit is variety of:

(a) Mango (b) Pomrgranate

(c) Guava (d) Grape

516. Most of annona species are belong to which climate?

(a) Tropical (b) Temperate

(c) Arid (d) All of above

517. Seedless cultivar of mandarin is:

(a) Nagpur (b) Khasi

(c) Coorge **(d) Satsuma**

518. Arka Trishna is important cultivar of:

(a) Guava **(b) Grape**

(c) Pomegranate (d) Aonla

519. Fruit cracking in litchi is due to deficiency of:

(a) Ca (b) Zn

(c) B (d) K

520. Caprification is associated to which crop?
(a) Mango **(b) Fig**
(c) Grape (d) Litchi

521. Ber takes how much days for maturity?
(a) 110-120 days (b) 120-150 days
(c) 150-180 days (d) 180-200 days

522. Propagation method of loquat is:
(a) Air layering (b) Patch budding
(c) Tongue grafting (d) T-budding

523. MM series rootstock of apple is cross of Mailing series with:
(a) Crab apple **(b) Northern Spy**
(c) EMLA series (d) None

524. Pusa Early Dwarf is very important cultivar of:
(a) Strawberry (b) Tomato
(c) Brinjal (d) Kiwi fruit

525. Day neutral variety of strawberry is:
(a) Selva (b) Tristar
(c) Tribute **(d) All of above**

526. How many month old coconut seedling is planted generally?
(a) 6-8 months **(b) 9-12 months**
(c) 12-13 months (d) 13-15 months

527. Guava wilt is mainly restricted to:
(a) Alkali soils (b) Acidic soils
(c) Alkaline soils (d) Neutral soils

528. Yellow leaf of citrus is caused due to deficiency of:
(a) Mo (b) Mg
(c) Zn (d) B

529. TOP (tapping panel dryness) in rubber caused due to:
(a) Water deficiency
(b) High temperature
(c) Excessive harvesting of latex
(d) None

530. Tea yellow disease of tea plants is due to deficiency of:
(a) Mg **(b) S**
(c) Mn (d) Mo

531. Cock's comb is physiological disorder of:
(a) Grapes (b) Fig
(c) Litchi **(d) Sapota**

532. Urigam is popular cultivar of:
(a) Tamarind (b) Coconut
(c) Drumstick (d) None

533. Which fruit is protogynous in nature?
(a) Banana (b) Fig
(c) Both a & b (d) Sapota

534. Which fruit is protandrous in nature?
(a) Banana (b) Fig
(c) Both a & b **(d) Sapota**

535. Mitha Chikna and Mithotra are important cultivars of:
(a) Sweet lime (b) Acid lime
(c) Sweet orange (d) Grape fruit

536. Chromosome no. of European plum is:
(a) 48 (b) 16
(c) 32 (d) 18

537. Which fruit is Auto-hexaploid?

(a) Strawberry **(b) Kiwi**

(c) Persimmon (d) Both a & b

538. Which of following is known as 'King of spices?

(a) Large cardamom **(b) Black pepper**

(c) Small cardamom (d) Cinnamon

539. Which fruit is known as 'Fancy fruit'?

(a) Kiwi fruit (b) Papaya

(c) Mandarin (d) Sapota

540. Common name of Jamun is:

(a) Indian blackberry (b) Indian gooseberry

(c) Chinese gooseberry (d) Chinese blackberry

541. Which one of following is climatric fruit?

(a) Fig (b) Ber

(c) Citrus (d) Pineapple

542. Fig is originated from which place?

(a) Europe (b) Brazil

(c) Asia Minor **(d) West Asia**

543. Which one fruit is known as "Star fruit"?

(a) Phalsa **(b) Carambola**

(c) Apple (d) Banana

544. Which one fruit is known as "Star apple"?

(a) Phalsa (b) Carambola

(c) Apple (d) Banana

545. Average yield of papaya is (t/hec) :

(a) 16 (b) 30

(c) 60-75 (d) 80

546. India ranks in the production of mango is:

(a) 1st (b) 2nd

(c) 3rd (d) 5th

547. Edible part of fig is:

(a) Fleshy thalamus **(b) Fleshy receptacle**

(c) Fleshy placenta (d) Paricarp

548. Common name of Aonla is:

(a) Indian blackberry **(b) Indian gooseberry**

(c) Chinese gooseberry (d) Chinese blackberry

549. Which cultivar of mango has high fruit retention?

(a) Rumani (b) Chausa

(c) Deshehari (d) Langra

550. No. of plants of Amarpalli under HDP is:

(a) 400 **(b) 1600**

(c) 3000 (d) 6400

551. Mutant of cultivar Nendran is:

(a) Monthan (b) Rasthali

(c) Plantain (d) Moongli

552. Which one is gynodioecious variety of papaya is:

(a) Pusa Nanha (b) Pusa Dwarf

(c) Pusa Delicious (d) Pusa Giant

553. CO-1 variety of sapota is bred through cross of:

(a) Cricket Ball x Oval (b) Guthi x Oval

(c) Cricket Ball x Valivalsa (d) Guthi x Kirti Bharti

554. Sweet lime is native to which place?

(a) China **(b) India**

(c) Brazil (d) Japan

555. Which fruit is known as "Breakfast fruit"?

(a) Jackfruit (b) Banana
(c) Grape **(d) Grape fruit**

556. Cheapest training system in grapes is:

(a) Head (b) Bower
(c) Kniffin (d) Telephone

557. Common name of kiwi is:

(a) Indian blackberry (b) Indian gooseberry
(c) Chinese gooseberry (d) Chinese blackberry

558. Grape rootstock dogridge is resistant to:

(a) Nematode (b) Salt
(c) Phylloxera **(d) All of above**

559. Ruby is popular cultivar of:

(a) Guava **(b) Pomegranate**
(c) Grape fruit (d) All of above

560. Neelum is popular cultivar of:

(a) Grape **(b) Aonla**
(c) Apple (d) Ber

561. Percentage of male plants required for proper pollination in date palm is:

(a) 5% **(b) 10%**
(c) 20% (d) 33%

562. Propagation in kiwi fruit is done by:

(a) Hardwood cutting (b) Seed
(c) Softwood cutting (d) Patch budding

563. Fermented beverage cidar is produced from:

(a) Apple (b) Pear
(c) Date (d) Cherry

564. Which variety of grape is commercially cultivated in north India?
(a) Thompson Seedless (b) **Perlette**
(c) Black Champa (d) Sharad Seedless

565. State having highest productivity of guava is:
(a) Andhra Pradesh (b) Karnataka
(c) Maharashtra (d) **Uttar Pradesh**

566. Origin centre of custard apple is:
(a) China (b) India
(c) Tropical America (d) Asia Minor

567. Witches broom is the major disorder of which crop?
(a) Cocoa (b) Coffee
(c) Tea (d) Coconut

568. Which cultivar of pear has highest TSS?
(a) Punjab Nectar **(b) Punjab Gold**
(c) Red Blush (d) Jargonella

569. Double Century is the variety of:
(a) Bael (b) Cocoa
(c) Coconut (d) Coffee

570. Indigenous cultivar of arecanut is:
(a) Mangla (b) Sumangla
(c) Sreemangla **(d) Mohitnagar**

571. Vertical continuous stem of cocoa is called as:
(a) Fan (b) Jorquette
(c) Chupan (d) None

572. Sex form of fig is:
(a) Dioecious (b) Monoecious
(c) Gynodioecious (d) Andromonoecious

573. Kiwi is propagated by:
(a) **Cutting** (b) Layering
(c) Grafting (d) Budding

574. Kinnow was introduced in India in:
(a) 1947 (b) 1949
(c) 1956 (d) **1959**

575. Early maturing mango variety of south India is
(a) Neelum (b) **Bengainpali**
(c) Alphonso (d) Dashehari

576. Jhumka or clustering in mango is due to:
(a) **Low temperature** (b) High temperature
(c) Water imbalance (d) None

577. Water requirement of grapes is high during:
(a) Flower formation (b) Vegetative growth
(c) **Berry formation** (d) Pruning

578. Muttam Varika is the variety of:
(a) Guava (b) Bael
(c) Passion fruit (d) **Jackfruit**

579. Whip grafting is modification of.
(a) Veneer grafting (b) **Tongue grafting**
(c) Inarching (d) Bridge grafting

580. Bael is which kind of fruit tree
(a) **Deciduous tree** (b) Evergreen tree
(c) Deciduous shrub (d) Evergreen shrub

581. Which fig is able to produce pollens?
(a) San pandro (b) Smyrna
(c) **Capri fig** (d) All of above

582. Girdling is most common in:

(a) Litchi (b) Citrus

(c) Grape (d) Papaya

583. Inflorescence of cashew nut is:

(a) Panicle (b) Fasicle

(c) Polygamonoecious (d) Andromonecious

584. Chintamani-1 is the variety of:

(a) Cashew (b) Cocoa

(c) Coffee (d) Coconut

585. Who is known as 'Father of Pomology?

(a) H.B. Frost (b) H Higgin

(c) De Candole (d) B.P.Pal

586. Major handicap in breeding of pineapple is:

(a) Self sterlity **(b) Self incompatibility**

(c) Heterostyly (d) Self pollination

587. Popular clone used as male parent in coffee breeding:

(a) Cauvery (b) Singara

(c) Athrey **(d) Taferikela**

588. Calyptra is common found in:

(a) Grape (b) Coffee

(c) Papaya (d) Apple

589. Chicken tongue is physiological disorder of:

(a) Grape (b) Citrus

(c) Litchi (d) Mango

590. Breba is parthenocarpically of

(a) Litchi (b) Grape

(c) Papaya **(d) Fig**

591. Which variety of mango has characteristic terpine flavour?

(a) Bombay Green **(b) Langra**

(c) Alphonso (d) Totopuri

592. Ultra dwarf root stock of apple is:

(a) M-9 (b) M-13

(c) M-27 (d) MM-104

593. Dwarf root stock of apple is:

(a) M-9 (b) M-13

(c) M-27 (d) MM-104

594. Apple scab is caused by:

(a) Fungi (b) Bacteria

(c) Virus (d) Algae

595. Which temperate fruit has lowest chilling requirement?

(a) Pear **(b) Peach**

(c) Plum (d) Cherry

596. In Papaya, cool climate cause:

(a) Small fruit size (b) Low seed/fruit

(c) Low TSS (d) Colour remains green

597. Duodichogamy is observed in which fruit crop?

(a) Avocado (b) Pistachionut

(c) Chestnut (d) Date

598. Chimerically transgenics was first exploited in which fruit crop?

(a) Papaya (b) Apple

(c) Orange (d) Mango

599. Iron deficiency is first show in which leaves?

(a) Young leaves (b) Old leaves

(c) Both (d) None

600. Pusa Srijan is a

(a) Nullisomic (b) Monosomlc

(c) Trisomic **(d) Tetrasomic**

601. Pusa Srijan, dwarfing rootstock of guava is recommended for which cultivar?

(a) Allahabad Safeda (b) Sardar

(c) Apple Colour (d) All

602. In Papaya, cool climate cause:

(a) Maleness (b) Femaleness

(c) Hermaphroditeness (d) Colour remains green

603. In Papaya, hot climate cause:

(a) Maleness **(b) Femaleness**

(c) Hermaphroditeness (d) Colour remains green

604. Recalcitrant seeds are common in

(a) Vegetables (b) Flowers

(c) Fruit trees (d) Cereals

605. Bottom heating technique is useful in propagation of:

(a) Mango (b) Bougainvillea

(c) Both a & b (d) Rose

606. Removal of side shoots in banana is known as:

(a) Mattocking (b) Deshooting

(c) Desukering (d) Dehorning

607. Udayam is new cultivar of:

(a) Mango (b) Guava

(c) Papaya **(d) Banana**

608. Production of fruits remained how much (approximation in MT) during year 2012-13 according to NHB?

(a) 55 (b) 68

(c) 77 (d) 84

609. Jelly less than 65% sugars required:

(a) Pasteurization (b) Sterilization

(c) Fermentation (d) Crystallization

610. Which one of following fruit belongs to cacti family also known as 'pitaya'?

(a) Dragon fruit (b) Passion fruit

(c) Bread fruit (d) Jack fruit

611. International Institute of Horticulture (IIH) is situated at:

(a) Peru (b) China

(c) Brazil (d) France

612. Headquarter of IB (former name IBPGR) is situated in which country?

(a) Brazil (b) India

(c) Philippines **(d) Italy**

613. Central Institute of Horticulture (CIH) is situated in which Indian state?

(a) Delhi **(b) Nagaland**

(c) AP (d) MH

614. Pollination in litchi is govern by:

(a) Autogamy (b) Water

(c) Wind **(d) Insects**

615. Sweetest variety of mango is:

(a) Chausa (b) Langra

(c) Alphonso (d) Bombay Green

616. Cultivar of banana suitable for chips making:

(a) Poovan (b) Robusta

(c) Nendran (d) Hill Banana

617. Arka Prabhat is new cultivar of:

(a) Mango **(b) Papaya**
(c) Guava (d) Grape

618. Apple colour is variety of which fruit crop?

(a) Mango (b) Papaya
(c) Guava (d) Apple

619. Arka Sahan cultivar of custard apple is cross of which atemoya species?

(a) Atemoya x squamosa (b) *Squamosa* x *atemoya*
(c) *Squamosa* x *muricata* (d) *Atemoya* x *chertmoya*

620. Most effective weedicide in the nursery is:

(a) Glyoposate **(b) Bromocil**
(c) 2-4 D (d) Isoprotron

621. Canker resistant cultivar of acid lime is:

(a) Pramalini (b) Jai Devi
(c) PKM-1 (d) None

622. Which one is self incompatible lemon variety?

(a) Eureka (b) Lisbon
(c) Pant Lemon-1 (d) Kagzi kalan

623. Largest producer of litchi is:

(a) Brazil (b) India
(c) Philippines **(d) China**

624. Fig is preserved in form of:

(a) Pickle **(b) Dry slices**
(c) Canned (d) Juice

625. Which variety of ber shows very high pollen sterility?

(a) Illachi (b) Gola
(c) Seb (d) Umran

626. Loquat is available in market is:

(a) **Jan-Feb** (b) March-May

(c) June-July (d) All of abovc

627. Maximum fruit setting in case of apple is:

(a) 1% **(b) 5%**

(c) 10% (d) 20%

628. Which apple species is resistant to apple scab?

(a) *M. floribunda* (b) *M. baccata*

(c) *M. domestica* (d) None

629. Mutant dwarf variety of cherry is

(a) Lambert **(b) Compact Stella**

(c) Sun Brust (d) Stella

630. Chakrata, Roopa, Karna are famous varieties of:

(a) Almond (b) Apple

(c) Walnut (d) Strawberry

631. Chandler and Tioga are cultivar of:

(a) Peach (b) Plum

(c) Walnut **(d) Strawberry**

632. Edible part of coconut is:

(a) Endosperm (b) Mesocarp

(c) Parianth (d) Bracts

633. Tea yellow disease is due to deficiency of:

(a) S (b) Mg

(c) Mn (d) B

634. Bunchy top of banana is transmitted by:

(a) White fly **(b) Aphid**

(c) Rhizome weevil (d) None

635. Edible part of citrus fruits is:

(a) Fleshy placenta (b)Juicy fasicle

(c) Juicy placental hairs (d) Paricarp

636. Which cultivar of mango has lowest no. of perfect flowers?

(a) Rumani (b) Langra

(c) Dashehari (d) Chausa

637. Under HDP Amarpalli is planted under which system?

(a) Square **(b) Triangular**

(c) Hexagonal (d) Meadow

638. Planting material of banana is:

(a) Suckers (b) Bits

(c) Peepers **(d) All of above**

639. Banana cultivar resistant to cold

(a) Rasthali (b) Nendran

(c) Rajapuri (d) Monthan

640. Scientific name of mountain papaya is:

(a) ***C. candamarcensis*** (b) *C. carandas*

(c) *C. quertifolia* (d) *C. rugosa*

641. Sweetest and best flavoured fruit cultivar of pineapple is:

(a) Kew **(b) Sugar Loaf**

(c) Carbenzona (d) Mauritius

642. Chatri is important cultivar of:

(a) Peach (b) Plum

(c) Sapota (d) Strawberry

643. Rudrakshi is popular variety of:

(a) Jackfruit (b) Grape

(c) Bael (d) None

644. Which one is triploid citrus fruit?

(a) Pummelo (b) Grape Fruit

(c) Lime **(d) Tahiti Lime**

645. Which citrus fruit is Fruit of the New World?

(a) Pummelo **(b) Grape fruit**

(c) Mandarin (d) Sweet orange

646. Which system of training in grapes has high economic ratio?

(a) Bower (b) Head

(c) Kniffin (d) Telephone

647. Temple rootstock of grape is resistant to:

(a) Nematode (b) Salt

(c) Drought **(d) Pierce's disease**

648. Multiple cross variety of pomegranate is:

(a) Ganesh (b) Jyoti

(c) Mridula **(d) Ruby**

649. Aonla variety used for fruit canning is:

(a) Banarasi **(b) Neelam**

(c) Chakiya (d) Francis

650. Which fruit tree has feet in running water & head in the fire of sky?

(a) Coconut (b) Oil palm

(c) Date palm (d) Ber

651. Best pollinizer cultivar for Heyward cultivar of kiwi is:

(a) Tomuri (b) Bruno

(c) Monty (d) Abbotte

652. Ambrich is cross of:

(a) Ambri x Rich-a Red **(b) Rich-a-Red x Ambri**

(c) Ambri x Red Delicious (d) Red Delicious x Ambri

653. Best time of pruning in the peach is:

(a) Mach-Apr (b) June-July

(c) Oct-Nov **(d) Dec-Jan**

654. Stanley and President are important cultivars of:

(a) Pear (b) Peach

(c) Plum (d) Apple

655. UN-50 is cultivar of:

(a) Pea **(b) Cashew**

(c) Onion (d) None

656. First step in processing of tea is:

(a) Withering (b) Fermentation

(c) Skiffing (d) Tipping

657. Kent is important variety of:

(a) Tea (b) Cocoa

(c) Coffee (d) Coconut

658. Betel vine is trained on which training system?

(a) Bareja (b) T-bar

(c) Pergola (d) Knifftn

659. Sunheri variety of apple is cross of:

(a) Ambri x Golden Delicious (b) Golden Delicious x Ambri

(c) Ambri x Red Delicious (c) None

660. Self sterile variety of pear is:

(a) Bartlett (b) Thames Pride

(c) Flemish Beauty (d) Leconte

661. Neemaguard resistant rootstock to nematodes is popular rootstock of:

(a) Pear **(b) Peach**

(c) Plum (d) Apricot

662. Cultivar of plum widely cultivated in lower hills of UP is:

(a) Kalsey (b) Frontier

(c) Elephant Heart **(d) Satluj Purple**

663. Chaubattia Princess & Chaubattia Anupam is important cultivar of:

(a) Peach (b) Plum

(c) Apricot **(d) Apple**

664. Chaubattia Alankar is important cultivar of:

(a) Apple **(b) Apricot**

(c) Pear (d) Peach

665. Ullal series of cashew belongs to which state?

(a) Kerala **(b) Karnataka**

(c) Maharashtra (d) Tamilnadu

666. Plucking of recovery bushes first time is called:

(a) Skiffing (b) Skufifing

(c) Withering **(d) Tipping**

667. HDT (Hybrido-de-Timor) is related to which crop?

(a) Tea (b) Cocoa

(c) Coffee (d) Coconut

668. Crown gall formation in apple is caused by :

(a) Bacteria (b) Fungus

(c) Virus (d) Algae

669. Dry neck is physiological disorder of:

(a) Persimmon (b) Apple

(c) Avocado (d) Sapota

670. Which one of following is not an aggregate fruit?

(a) Raspberry (b) Strawberry

(c) Custard apple **(d) Fig**

671. Palmyra palm is native to which place?

(a) India **(b) S.E. Asia**

(c) Africa (d) Brazil

672. Crown choking in banana is caused due to:

(a) Low temperature (b) High temperature

(c) B deficiency (d) Potassium deficiency

673. Crown chocking in coconut is caused due to

(a) Low temperature (b) High temperature

(c) B deficiency (d) K deficiency

674. Calyx end rot in persimmon is caused due to deficiency of:

(a) Ca (b) B

(c) Cu (d) Zn

675. Wooliness is physiological disorder of:

(a) Apple (b) Pear

(c) Peach (d) Plum

676. Hazelnut belongs to which family?

(a) Fagaceae (b) Saxiferaceae

(c) Bombaceae **(d) Betulaceae**

677. Short styled pisitillate flowers found in:

(a) Common fig **(b) Capri fig**

(c) Smyrna fig (d) San padro fig

678. Highest samples are taken in which fruit crop for sampling?

(a) Mango (b) Guava

(c) Banana **(d) Grape**

679. Latex collected from the papaya is dried at what temperature?

(a) 0°C (b) 27°C

(c) 55°C (d) 90°C

680. Inflorescence of fig is:

(a) **Hypanthodium** (b) Spathe

(c) Corymb (d) Catkin

681. Type of self incompatibility in ber is:

(a) Homomorphic (b) Sporophytic

(c) **Gametophytic** (d) None

682. Acid lime is indicator of:

(a) Greening (b) **Canker**

(c) Tristeza virus (d) Gummosis

683. Peach is native to which place?

(a) Europe (b) India

(c) Brazil (d) **China**

684. Mango has which type of apomixis?

(a) **Adventitious** (b) Recurrent

(c) Non recurrent (d) Nuceller budding

685. Minimum number of male plants in the apple orchard is:

(a) 5% (b) 20%

(c) **33%** (d) 50%

686. Which mango cultivar has taurpine flavour?

(a) Chausa (b) Bombay Green

(c) Alphonso (d) **Langra**

687. Banana cultivar suitable for jam making is:

(a) Poovan (b) **Hill Banana**

(c) Nendran (d) Robusta

688. CO-1 cultivar of papaya is breed through:

(a) **Selection** (b) Hybridization

(c) Mutation (d) Introduction

689. Kohir Safed is important cultivar of:

(a) Apple (b) Mango

(c) Guava (d) Banana

690. African Pride is one of very popular cultivar of:

(a) Guava **(b) Custard apple**

(c) Avocado (d) Kiwi fruit

691. Finest mandarin in the world is:

(a) Nagpur (b) Kinnow

(c) Satsuma (d) Khasi

692. Jai Devi is variety of:

(a) Sweet orange (b) Tahti lime

(c) Acid lime (d) Mandarin

693. Grape cultivar arkavati cross of:

(a) Black Champa x Thompson Seedless

(b) Banalore Blue x Black Champa

(c) Banalore Blue x Anab-e-Shahi

(d) Black Champa x Anab-e-Shahi

694. Flowers of litchi are devoid of:

(a) Calyx **(b) Corolla**

(c) Androecium (d) Gynoecium

695. Which fig has long styled pistillate flowers?

(a) Poona fig (b) San pendro

(c) Symrna fig **(d) Wild fig**

696. Goma Kirti cultivar of ber is selection from:

(a) Seb **(b) Umran**

(c) Gola (d) Illaichi

697. Which cultivar of loquat is popular in south India?

(a) **Golden Yellow** (b) Thames Pride

(c) Pale Yellow (d) Tanaka

698. Which fruit shows zygodormancy?

(a) Apple (b) Mango

(c) **Aonla** (d) Guava

699. Semi dwarf variety of cherry is:

(a) Compact Stella (b) **Compact Lambert**

(c) Stella (d) None

700. Selva and Premier are cultivars of:

(a) Kiwi fruit (b) Litchi

(c) Apple (d) **Strawberry**

701. Coconut husk or mesocarp has coir to fiber ratio is:

(a) **70:30** (b) 30:70

(c) 50:50 (d) 60:40

702. Citrus canker is transmitted by:

(a) Citrus psylla (b) Citrus aphid

(c) **Leaf-minor** (d) White fly

703. Exanthema in citrus is caused due to deficiency of:

(a) Zn (b) **Cu**

(c) B (d) Mo

704. Himsagar is cultivar of:

(a) Banana (b) Guava

(c) **Mango** (d) Papaya

705. Banana cultivar for culinary purpose is:

(a) Poovan (b) **Monthan**

(c) Nendran (d) Pey Kunnan

706. Most popular cultivar of mandarin in the south India is:

(a) Khasi (b) Nagpur

(c) Kinnow **(d) Coorg**

707. Chakradhar is important cultivar of:

(a) Acid lime (b) Sweet lime

(c) Mandarin (d) Sweet orange

708. Arka Chitra is famous cultivar of:

(a) Mango **(b) Grape**

(c) Guava (d) Aonla

709. What is correct about square system?

(a) Simplest & ubiquitous

(b) Having less trees than rectangular, hexagonal & diagonal system

(c) Known as quincunx system when filler is planted

(d) All of above

710. What is correct about hexagonal system?

(a) Also known as equilateral triangular system

(b) Offer 15% more plants/unit area than square system

(c) Adopted usually on fertile, expensive & resources rich land

(d) All of above

711. What is true about quincunx or diagonal system?

(a) Similar to square system except a filler plant

(b) Offer 1.5 times more plant/unit area than square system

(c) Fillers Crops should be quick growing early maturing & remunerative

(d) All of above

712. What is correct about contour and terrace system?

(a) Usual orchard on hills, slopy & undulated lands

(b) Check soil erosion & conserve soil moisture

(c) Terrace systemic extension of contour system

(d) All of above

713. Grape cultivar suitable for double cropping is:

(a) Arka Hans **(b) Arka Shyam**

(c) Arka Vati (d) Arka Neelmani

714. Litchi plant grows luxuriantly at what temperature?

(a) 18°C (b) 24°C

(c) 27°C **(d) 30°C**

715. Swarn Roopa is famous cultivar of:

(a) Mango (b) Grape fruit

(c) Litchi (d) Strawberry

716. Which one is intermediate fig group?

(a) Poona fig **(b) San pendro**

(c) Smyrna fig (d) Wild Capri fig

717. Which cultivar of ber is most suitable for humid region?

(a) Gola (b) Seb

(c) Umran **(d) Mehroon**

718. 'Irish Peach' is well known cultivar of:

(a) Apple (b) Pear

(c) Peach (d) Plum

719. Which fruit crop is known as "Dollar earning crop"?

(a) Coconut (b) Coffee

(c) Cashew (d) Apple

720. Propagation in lemon is done by:
(a) Seed (b) **T-budding**
(c) Inarching (d) Cuttings

721. Fruit bud differentiation in aonla is in months of:
(a) March-April (b) April-May
(c) May-June (d) June-July

722. Powdery mildew resistant variety of mango is:
(a) Langra (b) Lal Sindhuri
(c) Kesar (d) Fazli

723. Coconut water contains how much percentage of water?
(a) 70% (b) 90%
(c) 94.5% (d) 97%

724. Leaf mottling of citrus is caused due to deficiency of:
(a) Cu **(b) Zn**
(c) B (d) Mg

725. Hen & chicken is physiological disorder of:
(a) Litchi (b) Apple
(c) Grape (d) Sapota

726. Wood apple belongs to which family?
(a) Rutaceae (b) Rosaceae
(c) Anacardiaceae (d) Annonaceae

727. Which fruit crop is protogynous in nature?
(a) Banana (b) Fig
(c) Sapota **(d) Both a & b**

728. Pecan nut has which kind of dichogamy?
(a) Heterodichogamy (b) Protoandry
(c) Protogyny (d) PDSD

729. Avocado has which kind of dichogamy?
(a) Heterodichogamy
(b) Protoandry
(c) Protogyny
(d) PDSD

730. Which fruit is known as 'Queen of Fruits'?
(a) Mangosteen
(b) Avocado
(c) Kiwi
(d) Pistachionut

731. Which fruit is known as Indian gooseberry?
(a) Date
(b) Aonla
(c) Kiwi
(d) Jamun

732. Propagating material for pineapple is:
(a) Suckers
(b) Slips
(c) Both a & b
(d) Offsets

733. Propagating material of strawberry is:
(a) Sucker
(b) Rhizome
(c) Offset
(d) Runner

734. Fruit bud differentiation in grapes is in :
(a) March-April
(b) April-May
(c) May-June
(d) June-July

735. Which cultivar of most mango is most regular bearer & free from spongy tissue?
(a) Neelum
(b) Amarpalli
(c) Mallika
(d) Ratna

736. Desukering in banana is performed once in how much days?
(a) 20 days
(b) 30 days
(c) 45 days
(d) 60 days

737. Genetic constituent of Nendran is:
(a) AAA
(b) AAB
(c) ABB
(d) AABB

738. Which variety of papaya is known as 'Madhu Bindhu'?
(a) Sunrise Solo (b) Taiwan
(c) Coorge Honey Dew (d) Surya

739. Kinnow hybrid is bred through cross of :
(a) King x Willow Leaf (b) Willow Leaf x King
(c) King x Trifoliate Orange (d) Trifoliate Orange x King

740. Sister seedling of Kinnow is:
(a) Citrange **(b) Wilking**
(c) Tangerine (d) Tahiti lime

741. Which citrus fruit is also known as 'Persian apple?
(a) Citron (b) Sweet lime
(c) Pummelo (d) Grape fruit

742. Most winter hardy rootstock of citrus is:
(a) Trifoliate orange (b) Rough lemon
(c) Rangpur lime **(d) Troyer citrange**

743. Which mineral is universal deficient in grape production?
(a) Zn **(b) Mg**
(c) Fe (d) Cu

744. Perlette famous cultivar of grape in Northern India takes how much day for maturity?
(a) 90-95 days (b) 100-120 days
(c) 120-150 days (d) 150-180 days

745. Most suitable and best time for air layering in litchi is:
(a) January-March (b) April-May
(c) July-October (d) November-December

746. Pomegranate is propagated through:
(a) Hardwood cutting (b) Patch budding
(c) T-budding (d) Ring budding

747. Capri fig has which sex form?
(a) Gynomonecious (b) Dioecious
(c) Monoecious (d) Gynodioecios

748. Goma Ashwarya is new cultivar of:
(a) Mango (b) Grape
(c) Guava **(d) Aonla**

749. Caprification in fig is:
(a) Drying (b) Notching
(c) Pollination (d) Propagation

750. Time of leaf pruning in date palm is done in months of:
(a) March-April **(b) June**
(c) Sept-Oct (d) Dec

751. Ploidy level of Gola variety of Ber is:
(a) 2x (b) 6x
(c) 4x **(d) 8x**

752. Who produced the first apple cultivar?
(a) T.A. Knight (b) H. D. Olmo
(c) J. H. Higgin (d) H. B. Frost

753. Winter Banana is important variety of :
(a) Grape fruit **(b) Apple**
(c) Guava (d) Banana

754. Grit cells are main featres of which fruit plant?
(a) Apple (b) Peach
(c) Plum **(d) Pear**

755. Grit cells free variety of pear is?
(a) Magness **(b) a & c**
(c) Flemish Beauty (d) Punjab gold

756. Nectrine are single mutant fuzzless bud sport of which fruit?

(a) Peach (b) Plum

(c) Loquat (d) Apricot

757. Vengurla series of cashewnut is belongs to which state?

(a) Kerala **(b) Karnataka**

(c) Maharashtra (d) Tamilnadu

758. Common male parents being used in hybridization programme of coconut is:

(a) COD (b) Gangabondam

(c) CGD **(d) Both a & b**

759. First plucking in tea is done after how much time of planting?

(a) 10-12 months (b) 12-15 months

(c) 18-20 months (d) 24-28 months

760. Which one of following is multiple fruit?

(a) Fig (b) Strawberry

(c) Aonla (d) Pomegranate

761. Which of the following plant is day neutral plant?

(a) Papaya (b) Guava

(c) Both a & b (d) Cashew Nut

762. Carambola belongs to which family?

(a) Saxiferaceae **(b) Oxalidaceae**

(c) Guttiferaceae (d) Carricaceae

763. Strawberry is native to which place?

(a) China (b) Himalaya

(c) Brazil **(d) None**

764. Which one is known as "Food of God"?

(a) Coffee **(b) Cocoa**

(c) Bael (d) Banana

765. Grape is native to which place?

(a) **Black to Caspian sea** (b) Europe

(c) Mediterranean sea (d) S. E. Asia

766. Propagation in grape is done by:

(a) Softwood cutting (b) Patch budding

(c) **Hardwood cutting** (c) Ring budding

767. Black Corianth variety of grape has which kind of parthenocarpy?

(a) Stenospermocarpy (b) Vegetative

(c) **Stimulative** (d) None

768. Pollination in oil palm is done by

(a) Honey bee (b) Wasp

(c) Hand pollination (d) **Weevil**

769. Edible part of coconut is:

(a) Mesocarp (b) Endocarp

(c) **Endosperm** (d) Perianth

770. Mango seed kernel has percentage of protein:

(a) **9.5 %** (b) 22 %

(c) 24 % (d) 12 %

771. Akshay cultivar of mango is selection from:

(a) Langra (b) **Deshehari**

(c) Neelam (d) Ratna

772. Semi tall sport of Dwarf Cavendish banana is:

(a) Grand Naine (b) **Robusta**

(c) Ney Poovan (d) Nendran

773. Papaya cultivar suitable for papain extraction is:

(a) Pusa Delicious (b) **Pusa Majesty**

(c) CO-1 (d) Pusa Nanha

774. Scientific name of Chinese guava is:

(a) *P. fridrichsthalianum*
(b) *P. guajava*
(c) *P. cattelcianum*
(d) None

775. Most common rootstock of custard apple is:

(a) *A. squamosa*
(b) *A. reticulata*
(c) *A. atemoya*
(d) *A. muricata*

776. Which one is non-astringent variety of persimmon is?

(a) Fiyu
(b) Triumph
(c) Hachiya
(d) Nightingale

777. Which one is astringent variety of persimmon is?

(a) Fiyu
(b) c & d
(c) Hachiya
(d) Nightingale

778. Which rootstock citrus is resistant to greening?

(a) Rangpur lime
(b) Adjamir
(c) Troyer citrange
(d) Flying Dragon

779. Which rootstock citrus is resistant to tristeza virus?

(a) Rangpur lime
(b) Adjamir
(c) Rough lemon
(d) Flying Dragon

780. Who has remained famous grape breeder?

(a) J.H. Higgin
(b) H.D Olmo
(c) H.B. Frost
(d) None

781. Propagation of litchi is done by:

(a) Gooti
(b) Seed
(c) Softwood grafting
(d) Inarching

782. Pruning in fig is done in month of:

(a) March
(b) June
(c) October
(d) December

783. Avocado is botanically which kind of fruit?

(a) Berry (b) Nut

(c) Drupe (d) Capsule

784. Powdery mildew resistant variety of ber is:

(a) Sanur-2 (b) Dodhia

(c) Tikdi (d) Umran

785. Seedless cultivar of jamun is:

(a) Raj jamun (b) Paras

(c) Narendra Jamun-6 (d) Desi jamun

786. Thames Pride is popular cultivar of:

(a) Peach **(b) Loquat**

(c) Plum (d) Apricot

787. Most winter hardy rootstock of apple is:

(a) MM-104 (b) M-27

(c) MM-111 (d) Merton- 793

788. Seed stratification process takes how much days to break apple seed dormancy?

(a) 30 days (b) 30-60 days

(c) 60-90 days (d) 90-120 days

789. Which cultivar of pear is suitable to grow in South Indian conditions?

(a) Leconte (b) Sand Pear

(c) Gola **(d) Kiffer**

790. Which training system is most suitable for peaches?

(a) Modified central leader system

(b) Modified leader system

(c) Open central leader system

(d) Central leader system

791. Most common plum varieties grown in India are:

(a) Japanese (b) European

(c) Hybrids (d) None

792. Which plum cultivar reaches earliest in the Indian markets?

(a) Jamuni (b) Santa Rose

(c) Frontier **(d) Titron**

793. Which one is clonal rootstock of cherry?

(a) Pixy **(b) Colt**

(c) Paja (d) Mazzard

794. Most popular variety of almond is:

(a) Non-Pareil (b) Ne Plus Ultra

(c) Pearless **(d) All of these**

795. Maturity indices of walnut harvesting is:

(a) Peel or wall Cracking (b) Fruit turn yellow

(c) Fruit peel turn hard (d) None

796. Flower colour in strawberry is:

(a) White (b) Red

(c) Pink (d) Pale yellow

797. Coconut is botanically which kind of fruit?

(a) Nut (b) Berry

(c) Stone fruit (d) Capsule

798. Kurumba -an immature a refreshing clear liquid prepared from which kind of fruit?

(a) Coconut (b) Arecanut

(c) C. Nut (d) Date palm

799. Triploid variety of tea is:

(a) Sundram (b) Singara

(c) Jayaram (d) Athrey

800. Best quality cocoa is produced from:

(a) **Criollo** (b) Forestro

(c) Trinitarion (d) All of above

801. Parton dolly is dwarf cultivar of:

(a) Palmyra palm (b) **Oil palm**

(c) Coffee (d) Cocoa

802. Which cultivar is immune to panama wilt in banana?

(a) Gross Michel (b) Poovan

(c) **Basrai** (d) All of these

803. Leaf rust of coffee is caused by:

(a) *Gonoderma lucidum* (b) ***Hamelia vesatrix***

(c) *Tephrina deformens* (d) None

804. What is sprayed out to sort out problem of mango malformation?

(a) **NAA@ 200ppm** (b) Cultar

(c) IBA (d) 2-4 D

805. Fasciation is common physiological disorder in which fruit crop?

(a) Sapota (b) **Pineapple**

(c) Strawberry (d) Avocado

806. Kaveri is very popular cultivar of:

(a) Coconut (b) **Passion fruit**

(c) Tea (d) Coffee

807. Cauvery is a very popular cultivar of:

(a) Coconut (b) Passion fruit

(c) Tea (d) **Coffee**

808. Which one of following is only sub-tropical pome fruit?

(a) Pear (b) Peach

(c) Plum (d) **Loquat**

809. Rambutan belongs to which family?

(a) Sapindaceae (b) Sapotaceae

(c) Bombaceae (d) Betulaceae

810. Best material of propagation in pineapple is:

(a) Suckers (b) Slips

(c) Both a & b (d) Offsets

811. Propagation in ber is done by:

(a) Patch budding **(b) Ring or T-budding**

(c) Tongue grafting (d) Chip budding

812. Bud differentiation in pineapple is takes place at what stage?

(a) 10 leaves (b) 20 leaves

(c) 30 leaves **(d) 40 leaves**

813. Stone fruit is physiological disorder of:

(a) Sapota **(b) Custard apple**

(c) Avocado (d) Persimmon

814. Chestnut belongs to which family?

(a) Betulaceae **(b) Fagaceae**

(c) Bombaceae (d) Saxiferaccae

815. Chromosome no of guava is:

(a) 22 (b) 28

(c) 40 (d) 48

816. Stenospermocarpy requires which process for seedless fruit development?

(a) Pollination (b) Fertilization

(c) Both a and b (c) none

817. Which one is known as "king of spices"?

(a) Black pepper (b) Cardamom

(c) Clove (d) Cinnamon

818. Oldest known spice plant is:

(a) Cumin **(b) Cinnamon**

(c) Black pepper (d) Turmeric

819. Aonla is botanically which kind of fruit?

(a) Capsule (b) Berry

(c) Nut (d) Stone fruit

820. Which is short day plant (SDP)?

(a) Apple (b) Cashew nut

(c) Coffee **(d) Both b & c**

821. Botanical name of cucumber tree?

(a) Averrhoea bilmbi (b) *Actinidia deliciosa*

(c) *Feronia limonica* (d) None

822. Origin centre of sapota is:

(a) India (b) China

(c) Brazil **(d) Mexico**

823. Which fruit is commonly known as Chinese fig?

(a) Kiwi **(b) Ber**

(c) Aonla (d) Jamun

824. Propagation in pomegranate is done by

(a) Hardwood cutting (b) Softwood cutting

(c) Patch budding (d) T- budding

825. Sindhu seedless cultivar of mango has which kind of parthenocarpy?

(a) Vegetative (b) Stimulative

(c) Stenospermocarpy (d) None

826. Mode of pollination in banana is:

(a) Orniothophilus (b) Anemophilus

(c) Entomophilus (d) Self Pollination

827. Fruit bearing habit of guava is:

(a) Teminal **(b) Auxiliary**

(c) Mix bearing (d) None

828. Mango cultivar susceptible to mango malformation is:

(a) Mallika (b) Neelum

(c) Alphonso (d) Ratna

829. Inflorescence of guava is:

(a) Fasicle (b) Panicle

(c) Solitary (d) Spadix

830. Aril is edible part of which fruit?

(a) Litchi (b) Mango

(c) Strawberry (d) Guava

831. Salt tolerant & polyembryonic rootstock of mango is:

(a) Villaicolumban (b) Chandrakaran

(c) Kurukan (d) Bombay Green

832. Seedless cultivar of mango is:

(a) Ratna (b) Neelum

(c) Sindhu (d) Amarpalll

833. Genetic classification of banana is given by:

(a) Tanaka & Swingle

(b) Simmond & Shephard (1955)

(c) Bruns & Paryag (1911)

(d) Woodhouse

834. Banana is good source of:

(a) K (b) Ca

(c) Na (d) Fe

835. Most limiting factor for papaya cultivation is:

(a) Heat (b) Wind

(c) Frost (d) Water

836. Charlotte Rothchild is variety of:

(a) Peach (b) Plum

(c) Almond **(d) Pineapple**

837. Which citrus fruit has non-acid juice?

(a) Sweet lime (b) Kagzi lime

(c) Lemon (d) Acid lime

838. Parent of Pusa Seedless cultivar of grapes is:

(a) Thomson Seedless (b) Anab-e-Shahi

(c) Sharad Seedless (d) Parlette

839. What is marcottage or gootee in Litchi?

(a) Germination (b) Harvesting

(c) Air layering (d) Prunning

840. Highest fiber content among the fruits is contain in:

(a) Guava (b) Bael

(c) Mango **(d) Fig**

841. Most leading cultivar of avocado in the world is:

(a) Gotfried **(b) Furette**

(c) Taylor (d) None

842. Ber variety resistant to fruit fly is:

(a) Tikdi (b) Gola

(c) Dodhia (d) Seb

843. Which fruit seed is belong to recalcitrant category?

(a) Jamun (b) Ber

(c) Citrus **(d) All of above**

844. Golden yellow is famous cultivar of:
(a) Pear (b) Peach
(c) Loquat (d) Plum

845. East Mailing Long Aston (EMLA) series of apple rootstock resistant to:
(a) Wooly apple aphid (b) San jose scale
(c) Virus (d) All of above

846. Which temperate fruit come earliest in the market?
(a) Cherry (b) Apple
(c) Plum (d) Peach

847. Chakrata is very important cultivar of:
(a) Apricot (b) Almond
(c) Plum **(d) Walnut**

848. Commercial propagation in strawberry done by:
(a) Stolons **(b) Runners**
(c) Offsets (d) Suckers

849. Optimum temperature for growth of coconut is:
(a) 18°C (b) 23°C
(c) 27°C (d) 32°C

850. Singara is famous cultivar of:
(a) Tea (b) Coffee
(c) Coconut (d) Arecanut

851. Fruit of coffee is botanically:
(a) Berry **(b) Drupe**
(c) Nut (d) Capsule

852. Which cultivar of banana is resistant to panama wilt?
(a) Gross Michel (b) Basrai
(c) Poovan (d) None

853. Coffee species resistant to coffee rust is:

(a) *Coffea arabica* (b) *Coffea robusta*

(c) *Coffea canephora* **(d) *Coffea liberica***

854. Mango cultivar which is more prone to alternate bearing is?

(a) Alphonso (b) Bombay Green

(c) Neelum (d) Ratna

855. Sun Scald is physiological disorder of:

(a) Pineapple (b) Sapota

(c) Guava (d) Banana

856. Kaveri is popular cultivar of passion fruit is cross between groups:

(a) Purple x Green **(b) Purple x Yellow**

(c) Yellow x Purple (d) Yellow x Green

857. Which one of following fruit belongs to sub-tropical climate?

(a) Banana (b) Papaya

(c) Date Palm (d) Grape

858. Which fruit is native to Peru?

(a) Guava (b) Sapota

(c) Pineapple (d) Jackfruit

859. Which climateric fruit shows triploid sigmoid curve?

(a) Kiwi (b) Sapota

(c) Banana (d) Fig

860. Which one fruit is called as "Poor man's food"?

(a) Ber (b) Coconut

(c) Jackfruit (d) Date

861. Average yield of mango (in tons/hac) :

(a) 8 (b) 16

(c) 20 (d) 25

862. Tautra Trellis, HDP system is followed in which temperate fruit?

(a) **Peach** (b) Plum

(c) Apple (d) Loquat

863. Frontier is very popular cultivar of:

(a) Apple (b) Peach

(c) Plum (d) Apricot

864. Self fruitful variety of almond Is:

(a) Drake (b) Katha

(c) Dhebar **(d) All of above**

865. Bitter taste of tea is due to:

(a) Tannin (b) Phenolic compounds

(c) Thein (d) Theaflavin

866. Propagation in most of palms is by:

(a) Sucker (b) Offset

(c) Seed (d) Runner

867. Major devastating disease of papaya is:

(a) Damping off (b) **Ring spot virus**

(c) Leaf curling (d) Little Leaf

868. Major pest problem of sapota is:

(a) Moth (b) Mite

(c) White fly (d) Fruit & shoot borer

869. Pomegranate cultivar resistant to fruit cracking is:

(a) Khandari (b) Alandi

(c) Bedana Bosec **(d) All of above**

870. Family of filbert is:

(a) Betulaceae (b) Fagaceae

(c) Bombaceae (d) Saxiferaceae

871. Brown Turkey is important cultivar of:
(a) Peach (b) Plum
(c) Avocado **(d) Fig**

872. Which one is called as "Salad Fruit"?
(a) Kiwi (b) Passion fruit
(c) Avocado (d) Pomegranate

873. Dwarfing rootstock of ber suitable for HDP is:
(a) *Z. mauritiana* (b) ***Z. nummularia***
(c) *Z. jujuba* (d) *Z. rostra*

874. Late season cultivate loquat is:
(a) California Advance (b) Fireball
(c) Thames Pride (d) Safed

875. Commercial method of propagation of apple rootstock is:
(a) Seed **(b) Stooling**
(c) Air layering (d) Tongue grafting

876. Ultra high density is under which system of planting?
a) Contour system (b) Tatura trellis
(c) Triangular system **(d) Meadow system**

877. Stella is important variety of:
(a) Peach (b) Plum
(c) Cherry (d) Apple

878. Roopa is important cultivar of:
(a) Walnut (b) Peach
(c) Plum (d) None

879. Mulching is common cultural practice in which fruit crop?
(a) Kiwi **(b) Strawberry**
(c) Citrus (d) Litchi

880. Which group of coconut is shows autogamy?

(a) Dwarf (b) Tall

(c) Both (d) None

881. Fruit of oil palm is botanically known as:

(a) Nut (b) Berry

(c) Drupe (d) Stone

882. Origin centre of pineapple is:

(a) India (b) China

(c) Brazil (d) Europe

883. Carambola is rich in which acid?

(a) Tartaric acid **(b) Oxalic acid**

(c) Nicotinic acid (d) Citric acid

884. "Tree of Paradise" is common name of:

(a) Coconut (b) Date palm

(c) Banana (d) Apple

885. Storage temperature of pineapple is:

(a) 0°C (b) 5-6°C

(c) 11-13°C (d) 8-9°c

886. Largest producer of coconut is:

(a) India **(b) Indonesia**

(c) Sri Lanka (d) China

887. Inflorescence of ber is:

(a) Fasicle (b) Panicle

(c) Spadix (d) Solitary

888. Edible part of guava is:

(a) Thalamus (b) Pericarp

(c) Both a & b (d) Placenta

889. TSS of mango generally:

(a) 18 % **(b) 20 %**
(c) 22 % (d) 24 %

890. Sindu seedless cultivar of mango is developed by cross of:

(a) Ratna x Alphonso (b) Neelam x Alphonso
(c) Desahari x Neelum (d) Neelum x Desahari

891. Most bananas cultivar is in nature are:

(a) Diploid **(b) Triploid**
(c) Amphidiploid (d) Allotetraploid

892. Bodles Altafort, synthetic hybrid of banana is cross of:

(a) Gross Michel x Pisanglilin
(b) Bombay Green x Pisanglilin
(c) Gross Michel x Laden
(d) Bombay Green x Laden

893. Number of papaya plants under HDP system is around:

(a) 1600 (b) 3300
(c) 6000 (d) 6400

894. Variety of pineapple suitable for canning purpose is:

(a) Queen **(b) Kew**
(c) Mauritius (d) Jaldhup

895. Kirti Bharti is important cultivar of:

(a) Sapota (b) Litchi
(c) Peach (d) Plum

896. Propagation in jackfruit is done by:

(a) Seed (b) Patch budding
(c) Air layering **(d) Inarching**

897. Bud mutant of Nagpur mandarin is:

(a) Mosambi (b) Kinnow

(c) Mudkhed (d) Pineapple

898. Foster is well known cultivar of:

(a) Strawberry (b) Avocado

(c) Grape **(d) Grape fruit**

899. Back pruning or foundation pruning in grapes is practiced in months of:

(a) April (b) July

(c) September (d) December

900. Percentage of grapes used for wine making world-wide is:

(a) 10 % (b) 30 %

(c) 50 % **(d) 90 %**

901. A 20-30 years old chiku tree gives:

(a) 2000-3000 Fruits (b) 2,500-3,000 fruits

(c) 4,000-6,000 fruits (d) 700-800 fruits

902. A cashewnut tree will start fruiting after how many years of planting:

(a) 4 years (b) 9 years

(c) 7 years (d) 2 years

903. A chemical most effective in breaking bud dormancy in grapes:

(a) Dormex (Hydrogen cyanamide)

(b) GA3

(c) Thiourea

(d) NAA

904. A chemical used for artificial ripening in banana:

(a) Ethephon (b) Auxin

(c) KNO_3 **(d) Calcium carbide**

905. A coconut tree will start fruiting after how many years of planting:

(a) 2 years (b) **6 years**

(c) 1 year (d) 4 years

906. A cold-tolerant species of papaya:

(a) *C. papaya* (b) C. *candicana*

(c) *C. candamarcensis* (d) C. *cauliflora*

907. A common pest during flowering time of mango (Feb - March) is:

(a) Red ants (b) Shoot borer

(c) Mango Hopper (d) Mealy bug

908. A corm is an underground plant organ meant for:

(a) Storage of food (b) Asexual reproduction

(c) Both these (d) Sexual reproduction

909. A corn is a sudden base of:

(a) Stem axis (b) Leaf

(c) Leaf bud (d) Petiole

910. A disease which can transform a "On year'"of fruiting to "Off year" in mango is:

(a) Powdery mildew (b) Anthracnose

(c) Leaf blight (d) Stem canker

911. A dwarf root stock of apple is:

(a) M-9 (b) M-13

(c) HM-12 (d) M-25

912. A dwarf rootstock of pear is:

(a) Kaipth (b) Quince A

(c) Quince C (d) Quince

913. A fruit that develops from a hollow pear-shaped fleshy receptacle (hypanthodium) which encloses a number of minute male and female flowers is known as:

(a) Aggregate fruit (b) Hesperidium

(c) Pome fruit (d) **Syconus fruit**

914. A gene located on Y chromosome and therefore, transmitted from father to son is known as:

(a) Autosomal gene (b) Supplementary gene

(c) Sex linked gene (d) Holandric gene

915. A grafted mango tree yields about 2,000 to 5,000 fruits per year from:

(a) 5th year onward (b) 10th year onward

(c) 15th year onward **(d) 20th year onward**

916. A group of fruits which show neither a rise in respiration rate nor anassociated production of ethylene during the ripening process is called

(a) Parthenocarpic (b) Climacteric

(c) Non-climacteric (d) Parthenogenesis

917. A hectare guava plantation yields:

(a) 22,000 kg of fruits (b) 30,000 kg of fruits

(c) 10,200 kg of fruits **(d) 25,000 kg of fruits**

918. A hectare of grape plantation can yield up to:

(a) 50,000 kg fruits (b) 40,000 kg fruits

(c) 60,000 kg, fruits (d) 70,000 kg, fruits

919. A method of plant propagation in which adventitious roots are formed on stem while it is still attached to mother plant is called:

(a) Cutting (b) Suckering

(c) Budding **(d) Layering**

920. A part of a plant at the ground surface giving rise to new plants is called:

(a) Crown **(b) Sucker**

(c) Offshoot (d) Runner

921. A perfect jelly have pectin of:

(a) 1.5% **(b) 1.0%**

(c) 11.25% (d) 2.0%

922. A plants embryo is microscopic in size and globular shape in:

(a) Third stage of its development

(b) First stage of its development

(c) Second stage of its development

(d) None of these

923. A rhizome is meant for:

(a) Food storage (b) Propagation

(c) Both these (d) None of these

924. A rhizome may be:

(a) Above ground (b) Underground

(c) Both these (d) None these

925. A salt-tolerant rootstock for citrus:

(a) ***Severenia buxifolia*** (b) Trifoliate orange

(c) Rought lemon (d) Sour orange

926. A seedless grape hybrid developed at IIHR Bangalore is:

(a) Arkavati (b) Arka Hans

(c) Arka shyam (d) Single stalk system

927. A series of biochemical changes initiated by the catalytic production of ethylene making the changes from growth to sentence and Solving an increase in respiration rate is called:

(a) Maturity (b) Ripening

(c) Climatric (d) Senescence

928. A serious disease in ber which cause fruit drop is:

(a) Blue mold (b) Anthracnose

(c) Powdery mildew (d) Wilt

929. A serious physiological disorder mango orchards near brick kilns is:

(a) Spongy tissue (b) Black rot

(c) Black tip (d) Malformation

930. A single sigmoid growth curve is in fruit:

(a) Grape (b) Peach

(c) Apple (d) Guava

931. A spur-type apple cultivar is:

(a) Cox Orange Pippin **(b) Red spur**

(c) Golden Delicious (d) Rich-e-Red

932. Abortion of the embryo before maturation which results in seedlessness is a problem in improvement of which fruit crop?

(a) Banana (b) Mango

(c) Litchi **(d) Grape**

933. According to flowering habit date palm is:

(a) Monoecious (b) **Dioecious**

(c) Hermophrodite (d) Andromonoecious

934. According to fruit maturity time, Neelum is which season variety:

(a) Early (b) Mid

(c) Late (d) Very late

935. Acid lime-is susceptible to:

(a) Powdery Mildew (b) Pink Disease

(c) Bacterial Canker (d) Twing Blight

936. Acid lime (C. *aurantifolia*) is commonly called as:

(a) Lime (b) Lemon

(c) Kagzi lime (d) Rangpur lime

937. Acta Horticulture is a publication of.

(a) ASHS, USA (b) **ISHS, Belgium**

(c) ISHS, Japan (d) HIS, India

938. Addition of which one of the following leads to the red colour of jelly?

(a) Artificial colour (b) Pectin

(c) Acid (d) **Charring of sugar**

939. Adventitious roots may arise from:

(a) Aerial parts of plants
(b) Underground stems
(c) Relatively old roots
(d) All these

940. Adventitious shoots appear on.

(a) Roots
(b) Internodes of stems
(c) Both these
(d) None of these

941. African Pride, a Annona Variety is developed by cross between

(a) *Sugar apple x A. reticulata*
(b) *Sugar apple x A. Cherimoya*
(c) *Sugar apple x A. glabra*
(d) *Sugar apple x A. squomosa*

942. After harvesting cutting of banana pseudostem is known as:

(a) Desuckering
(b) Mattocking
(c) Propping
(d) Parring

943. Agricultural and Processed food products Export Development Authority (APEDA) was established at:

(a) 1980
(b) 1986
(c) 1985
(d) 1990

944. Agricultural Research Services (ARS) was started in:

(a) 1975
(b) 1980
(c) 1974
(d) 1970

945. Agricultural Scientists Recruitment Board (ASRB) was established in the year:

(a) 1970
(b) 1980
(c) 1975
(d) 1973

946. Air layering in practiced usually with:

(a) Grapes
(b) Persian lime
(c) Mango
(d) None of these

947. Air layering is most common in:

(a) Rose **(b) Litchi**

(c) Grape (d) Guava

948. All the commercial mandarin varieties grown in India belongs to species:

(a) *C. reticulata* (b) *C. limonia*

(c) *C. latifolia* (d) *C. paradisica*

949. Allahabad Safeda is the variety of

(a) Mango **(b) Guava**

(c) Loquat (d) Grapes

950. Allison and 'Tomuri' are varieties of Chinese gooseberry which bear peculiar flower type?

(a) Hermaphrodite flowers **(b) Male flowers**

(c) Neutral flowers (d) Female flowers

951. Almond belongs to family:

(a) Rosaceae (b) Rutaceae

(c) Moraceae (d) Lauraceae

952. Almond is originated from which centre of origin?

(a) Mediterranean centre (b) Chinese

(c) Asia Minor (d) Hindustan

953. Almond principally bears on the:

(a) Old shoots **(b) Spurs**

(c) New Shoots (d) Current season growth

954. Alternate bearing is most common is:

(a) Date palm (b) Apricot

(c) Cherry **(d) Mango**

955. Alternate bearing, a serious problem in mango is also seen in:

(a) Banana **(b) Apple**

(c) Citrus (d) None of these

956. Alternate forms of gene at the same locus are referred to as:

(a) Allis (b) Chromosome

(c) Plastid (d) Dominant

957. Ambebahar" and "Mrigh bahar" are varieties of:

(a) Sweet orange (b) Mandarin orange

(c) Mango (d) Banana

958. Amimgola is variety of.

(a) Mango (b) Fig

(c) Guava (d) Grape

959. Among the following pairs of grape varieties which are relatively salt resistant?

(a) Alamwick and Beauty Seedless

(b) Perlette and Khalili

(c) Anab-e Shahi & Bangalore Blue

(d) None of the above

960. Among the following which is the food preservator?

(a) Celphos (b) Calcium cynamide

(c) Sodium benzoate (d) Potassium permagnate

961. Amongst Annona species which bears the largest fruit?

(a) *Annona muricata* (b) *A. glabra*

(c) A. *cherimoia* (d) *A reticulata*

962. Amphimixis method of reproduction is:

(a) Asexual **(b) Vegetative**

(c) None of these (d) Sexual

963. Amrapali is a hybrid of which one of the following parents?

(a) Dashehari x Neelam (b) Prince x Fazli

(c) Gaurjeet x Himsagar (d) Alphanso x Ras puri

964. Amrit Sagar is a variety of:
(a) Mango **(b) Banana**
(c) Citrus (d) Guava

965. An adult plant produces sex cells in:
(a) All parts of its body **(b) Flowers**
(c) Roots (d) Leaves

966. An apple tree will give the fruits after how many years of planting?
(a) 5 years (b) 3 years
(c) 2 years (d) 8 years

967. An example of slow release fertilizer is:
(a) Coated urea (b) CAN
(c) DAP (d) KCI

968. An induced or artificial layering is performed by removing from stem
(a) Phloem (b) Cambium
(c) Both these (d) Xylem

969. An ovum is contained in:
(a) Stigma (b) Stamens
(c) Ovary (d) Style

970. Anab-e- roga is a serious disease of which of the fruit crop?
(a) Coconut (b) Rubber
(c) Cashewnut (d) Oil palm

971. Anab-e- roga is caused by:
(a) Fungus (b) Bacteria
(c) Virus (d) ML0's

972. Anab-e-Shahi a seeded grape cultivar was selected in 1930 from the collection of Nawab Baquer Ali Khan who was associated with its selection?
(a) Sh. RS Pilay (b) Nawab Baquer Khan
(c) Sh. Arve (d) None of these

973. Anab-e-Shahi is the most popular grape variety of:

(a) Punjab (b) **Andhra Pradesh**

(c) Tamil Nadu (d) Karnataka

974. Ancestor of *Ananas comosus* is:

(a) *A braccteatus* (b) *A erectifolius*

(c) *A. microstachys* (d) *A ananassoides*

975. Androdioecious form of sex is found in:

(a) Papaya (b) Fig

(c) Rambutan (d) None of these

976. Animal and plant cells can be differentiated by:

(a) Size

(b) Shape

(c) Conductivity

(d) Presence or absence of cell wall

977. Anola belongs to the family:

(a) Rosaceae **(b) Euphorbiaceae**

(c) Myrtaccae (d) Areceae

978. Anola is commercially grown in the state:

(a) UP (b) KN

(c) HP (d) MH

979. Anola is rich in:

(a) Vitamin A (b) Vitamin

(c) Vitamin B2 **(d) Vitamin C**

980. Anola variety suitable for processing is:

(a) Banarasi (b) Red tinged

(c) NA-4 (Kanchan) (d) Hathijhool

981. ANOVA' was defined by:

(a) AL Bowley (b) Horace Secrist

(c) Karl Pearson (d) **RA Fisher**

982. Anthers contain:

(a) Ova. (b) **Pollen grains**

(c) Both These (d) None of these

983. Aonla fruit is native to:

(a) China (b) Pakistan

(c) India (d) Srilanka

984. Aonla *Phyllanthuss embllica* (syn. *Embllica officinalis*) belongs to family:

(a) Rosaceae (b) **Euphorbiaceae**

(c) Rutaceae (d) Sapotaceae

985. Apomictic embryo is produced:

(a) Sexually (b) **Asexually**

(c) By both these ways (d) By none of these ways

986. Apomictic seedlings rootstock of apple:

(a) *Malus baccata* (b) *M. domestica*

(c) *M. sikkiminensis* (d) None of the above

987. Apomixis in flours is which of the following?

(a) Polyembryony (b) Recurrent

(c) Vegetative (d) Non-recurrent

988. Apomixis is known to occur in:

(a) Citrus (b) Onion

(c) Rasberry, apple **(d) All these**

989. Apple and cherry are:

(a) Cool season crops (b) Warm Season crops

(c) Arid zone crops (d) Semi-arid zone

990. Apple belongs to the family:
(a) Euphorbiaceae **(b) Rosaceae**
(c) Musaceae (d) Arecaeae

991. Apple colour development is more intense in which one of the following regions?
(a) Lower hills (b) High hills
(c) Mid hills **(d) Not affected by altitude**

992. Apple crop flowers during:
(a) December-January (b) September-October
(c) June-July **(d) March-April**

993. Apple hybrid developed as a result of a cross between Ambri x Golden Delicious of RFRES, Shalimar, J&K is:
(a) Sunehari (b) Lai Ambri
(c) Rich -e-Red (d) Maharajaji

994. Apple is a type of:
(a) Samara (b) Achene
(c) Caryopsis **(d) False fruit**

995. Apple is propagated by:
(a) Seed
(b) Shielding, tongue grafting
(c) Crown and side grafting
(d) Leaf cutting

996. Apple rootstock are easily propagated through:
(a) Layering (b) Seeds
(c) Hard wood cutting **(d) Stooling**

997. Apple scab is caused by:
(a) *Dematophora necatrix* **(b) *Venturia inaequalis***
(c) *Podoslhera leuchotricha* (d) None of these

998. Apple shows which of the following type of apomixis?

(a) Non recurrent **(b) Recurrent**

(c) Budding (d) Vegetative

999. Application of 200 ppm of ethral from September onwards induces flowering in:

(a) Apple (b) Loquat

(c) Mango (d) Banana

1000. Application of NAA (Planofix), Calcium carbide or Ethrel on pine apple plants at 45-leaf stage brings about:

(a) Defoliation **(b) Uniform flowering**

(c) Profuse rooting, (d) Profuse tillering

1001. Application of which chemical can control fruit drop in *Citrus sinensis*?

(a) NAA (b) IBA

(c) 2-4-5-T **(d) 2,4-D**

1002. Application of which chemical can suppress flowering in mango?

(a) Auxin **(b) Gibberellins**

(c) Kinetin (d) Paclobutrazol

1003. Application of which growth regulator can bring regularity in mango to some extent?

(a) Paclobutrazol (b) Auxin

(c) ABA (d) Ethephon

1004. Application of which of the following chemical acts as an aid for mechanical harvesting?

(a) GA_3 (b) IAA

(c) Ethephon **(d) ABA**

1005. Application of which of the following reduces mango malformation?

(a) Abscisic acid (b) GA_3 @ 10 ppm

(c) NAA @ 200 ppm (d) 2,4-D @ 20 ppm

1006. Approximate number of sapota plants in one hectare is:

(a) 220 **(b) 120**

(c) 320 (d) 400

1007. Apricot is shield-budded on root stocks of:

(a) Wild apricot or zardalu (b) Wild cherry

(c) MIX (d) Malling-merton

1008. Arecanut is commonly propagated by:

(a) Seed (b) Cuttings

(c) Grafting (d) Budding

1009. Aonla Fruit is a

(a) Capsule (b) Nut

(c) Pome (d) Berry

1010. Aril is edible portion of which of the following fruit crops?

(a) Apple, Pear, Karonda

(b) Mango, Loquat, Avocado

(c) Litchi, Pomegranate, Mangosteen

(d) Grape, Karonda, Phalsa

1011. Arka Amulya are the varieties of guava are developed through:

(a) Selection (b) Mutationc

(c) Hybridization (d) Polyploidy

1012. Arka Mridula are the varieties of guava are developed through:

(a) Selection (b) Mutationc

(c) Hybridization (d) Polyploidy

1013. Arka chitra belongs to:

(a) Pineapple (b) Apple

(c) Grape (d) Guava

1014. Arka Punneet is a hybrid between:

(a) Dashehari x Chausa (b) Dashehari x Neelum

(c) Amrapali x Janardhari Prasad **(d) Alphonso x Baganpalli**

1015. Arka Sahan is a commercial cultivar of

(a) Banana (b) Amla

(c) Custard apple (d) Pear

1016. Arka shyam is hybrid variety of:

(a) Jamun **(b) Grape**

(c) Mulberry (d) Guava

1017. Arkavati is a hybrid & Variety of:

(a) Banana **(b) Grape**

(c) Apple (d) Jamun

1018. Arkavati grape was developed at:

(a) IARI, New Delhi (b) PAU, Ludhiana

(c) IIHR, Bangalore (d) CITH, Srinagar

1019. Arkavati, seedless grape hybrid has been developed by a cross between:

(a) Bangalore Blue & Anab- e- Shahi

(b) Black Champa & T.S

(c) Bangalore Blue & Black Champa

(d) Anab-e-Shahi & Queen of vineyard

1020. Arka Sahan is a hybrid of Annona a developed as a result of cross between:

(a) *A. cherimoya x A. Squamosa*

(b) *A. reticulata x A. Squamosa*

(c) *A. Squamosa x A. glabra*

(d) *A. atemoya x A. Squamosa*

1021. Arka Aruna is cross between:

(a) Banganpalli x Alphonso

(b) Alphonso x Janardhan Pasand

(c) Alphonso x Banganpalli

(d) Alphonso x Dashahari

1022. African Pride is a hybrid of annona a developed as a result of cross between:

(a) *A. cherimoya x A. Squamosa*

(b) *A. reticulata x A. Squamosa*

(c) *A. Squamosa x A. glabra*

(d) *A. atemoya x A. Squamosa*

1023. Aroma in the ripe fruit of apple is due to:

(a) Ethyle-2 methyl butyrate (b) Eugenol

(c) Isopentanol (d) Hexanol

1024. Aroma of over ripe fruit of banana is due to:

(a) Diallyl propyl **(b) Isopentanol**

(c) Hexanol (d) Allyl propyl

1025. Around how much diameter are majority of coconut roots concentrated?

(a) 6 m (b) 2 m

(c) 8 m **(d) 4 m**

1026. Artificial hybridization in fruit crops in the world was first practiced by:

(a) Burns & Prayag (b) Tonaka

(c) Thomas Andrew Knight (d) Olmo

1027. Artificial ripening is very difficult in:

(a) Lemon (b) Papaya

(c) Persimmon (d) Plums

1028. Astringency in persimmon fruit is removed by:

(a) Ethylene **(b) GA**

(c) NAA (d) CK

1029. Astringency in persimmon is lost due to:

(a) Conversion of tannins to sugar

(b) Breakdown of mannitol to sugar

(c) Palymerization of tannins

(d) Breakdown of polysaccharides to sugars

1030. At present the maximum area under fruits in India is occupied by:

(a) Grapes (b) Guava

(c) Mango (d) Citrus

1031. At which spacing is planting in Arecanut done?

(a) 1.5 m (b) 5 m

(c) 2.7 m (d) 8 m

1032. Atmosphere extends above mean sea level to a height of about:

(a) 160 km (b) 1600 km

(c) 1600 miles (d) 1600 metres

1033. A triploid apple cultivar is:

(a) Ben Davis (b) Red Delicious

(c) Jonathan (d) Golden Delicious

1034. A mutant cultivar of mango is:

(a) Totapuri **(b) Rosica**

(c) Zafran (d) Himsagar

1035. Average temperature required for Banana is:

(a) 17 ^{0}C **(b) 27°C**

(c) 23°C (d) 35 ^{0}C

1036. AVG is a inhibitor of:

(a) Ethylene (b) Cytokinin

(c) Gibberellin (d) Auxin

1037. Avocado belongs to family:

(a) Rosaceae **(b) Lauraceae**

(c) Rutaceae (d) Moraceae

1038. Avocado contain high content of:

(a) Protein (b) Vitamin A

(c) Fat (d) Iron

1039. Avocado tress flower in February and April and their fruits are harvested in:

(a) Aug- Sep **(b) July**

(c) October (d) November

1040. *Bacillus Thruingenesis* (BT) is used in Sapota to control:

(a) Bud borer (b) Psylla

(c) Scale (d) Aphid

1041. Back crossing between which of the following resulted in the development of seedless mango ' Sindhu'?

(a) Ratna x Alphonso (b) Alphonso x Banganpalli

(c) Ratna x Neelum (d) None of these

1042. Backcross method of breeding means:

(a) Mutation breeding

(b) Tissue culture

(c) Polyploidy breeding

(d) Crossing back the F_1 to desirable parents

1043. Bacterial canker resistant hybrid of acid lime is:

(a) Hybrid (b) Hybrid 4

(c) NS 2 **(d) All of these**

1044. Bahar treatment in citrus is practiced in:

(a) East & North India (b) **Central & West**

(c) North & East India (d) West & South India

1045. Baldwin, triploid variety of apple is:

(a) Self unfruitful **(b) Self fruitful**

(c) Self sterile (d) None of these

1046. Banana belongs to the family

(a) Rhamnaceae (b) Annonaceae

(c) Musaceae (d) Vitaceae

1047. Banana genomic classification was given by:

(a) Simmond and Shepherd (b) De Candolle

(c) Deshmukh and Pandey (d) Linnoeaus

1048. Banana is generally planted in:

(a) Sept-Oct **(b) June-July**

(c) Feb-March (d) April-May

1049. Banana is grown mainly in Tamil Nadu, West Bengal, Kerala, Maharashtra, Gujarat, Karnataka, Assam, Andhra Pradesh and Bihar and occupies,

(a) Over 1,64 000 hectares (b) 1,00,000 hectares

(c) 80,000 hectares (d) 1,10,000 hectares

1050. Banana is propagated by means of suckers which arise

(a) Underground rhizomes **(b) Underground corms**

(c) Stolons (d) Pseudostems

1051. Banana planting time is:

(a) June-July (b) January

(c) December (d) October

1052. Banana should be propagated by:

(a) Sword suckers (b) Water suckers

(c) Rhizomes (d) Stem cutting

1053. Banana Streak Virus (BSV) is a serious problem and is transmitted through:

(a) Aphids (b) White fly

(c) Banana beetle (d) **Mealy bug**

1054. Banana variety which is immune to panama wilt is

(a) Rasthali **(b) Basrai**

(c) Lal Kela (d) Poovan

1055. Banana, papaya and jack fruits are used as:

(a) Fruits (b) Vegetables

(c) Both (d) Salad

1056. Barbados cherry *Malphigia puncifolia* (syn. *M. glabara*), richest source of vit. C. It belongs to family:

(a) Bombacaceae (b) Anacardiaceae

(c) Guttiferae (d) **Malphigiaceae**

1057. Based on mode of pollination Tall coconut cultivars are:

(a) Allogamous (b) Parthenocarpic

(c) Autogamous (d) None of these

1058. Based on the climatic requirements papaya is classified as:

(a) Temperate fruit **(b) Tropical fruit**

(c) Sub-tropical fruit (d) Sub-temperate fruit

1059. Basic chromosome number (x=) of mango:

(a) 20 (b) **10**

(c) 40 (d) 80

1060. Basic chromosome number of pomoideae is:

(a) 7 (b) **17**

(c) 9 (d) 8

1061. Basrai Dwarf is a variety of.

(a) Papaya **(b) Banana**

(c) Sweet orange (d) Mango

1062. Before planting in nursery grape cuttings are put in moist sand for about a month for:

(a) Conditioning (b) Activation

(c) Revitalization **(d) Callusing**

1063. Behat Coconut is seedless variety of guava. The Seedlessness is due to:

(a) Parthenocarpy (b) **Triploidy**

(c) Self incompatibility (d) Parthenogenesis

1064. Beneficial fungi associated with roots of litchi is:

(a) Phylophthora (b) Rhizobium

(b) Mycorrhizeae (d) Pythium

1065. Ber belong to the family:

(a) Myrtaceae (b) Rosaceae

(c) Rhamnaceae (d) Rutaceae

1066. Ber is commonly propagated by:

(a) Grafting (b) Stooling

(c) Ring budding (d) None of these

1067. Ber is native to:

(a) Tropical America

(b) China

(c) Malaysia

(d) **India and South West Asia**

1068. Ber plants are pruned in North India during:

(a) May-June (b) June-July

(c) April-May (d) July-Aug

1069. Ber variety 'Umran' a seedling selection from Rajasthan is commercially grown in:

(a) Punjab (b) **Madhya Pradesh**

(c) Rajasthan (d) Uttar Pradesh

1070. Berry elongation and thinning in grapes is done with:

(a) 2,4-D (b) Ethephon

(c) GA_3 (d) NAA

1071. Best time for the propagation of apple is:

(a) Dec-Jan (b) Feb-March

(c) April -May (d) July-Aug

1072. Best Method for storing citrus at room temperature for a few weeks:

(a) Submerge in cold water and dry

(b) Keep in well ventilated room

(c) Keep in atmospheric rich in CO_2

(d) Keep individual fruits in humidity proof envelops

1073. Best planting material of banana:

(a) Slips (b) Crown

(c) Suckers (d) Off-shoot

1074. Best planting material of pineapple is:

(a) Slip (b) Sucker

(c) Seed (d) Crown

1075. Best quality pomegranate fruits are produced in regions of

(a) Dry hot summer **(b) Prolonged winters**

(c) Cold rainy season (d) Wet hot summer

1076. Best rootstock for citrus in south India:

(a) Cleopatra mandarin **(b) Rangpur lime**

(c) Citrange (d) None of the above

1077. Best rootstock species for budding of many ber varieties is:

(a) *Z. rotundifolia* (b) *Z. mauritiana*

(c) *Z. rugosa* (d) *Z. xyloyrus*

1078. Best soil for planting of fruit orchard should have pH between

(a) 3.5-5.5 **(b) 6.5-7.5**

(c) 7.5-10.5 (d) 4.5-6.5

1079. Best time of budding in ber is:

(a) July-Aug (b) Sept-Oct

(c) May-June (d) Feb-March

1080. Best time of budding in peach, plum, pear and apple is:

(a) Jan-Feb (b) March-April

(c) May-June (d) July - August

1081. Best time of planting ber in arid regions is :

(a) July-August (b) Dec-Jan

(c) Feb-March (d) April- May

1082. Best time of pruning peach is:

(a) Mid winter (b) Spring

(c) Autumn (d) Mid summer

1083. Best time of transplanting loquat is:

(a) Jan-Feb **(b) Sept-Oct**

(c) Feb-March (d) March - April

1084. Best vegetative material for the propagation of pineapple is:

(a) Shoot suckers **(b) Ground suckers**

(c) Crown (d) Stumps

1085. Bhadauran, lllaichi and Alib varieties of mango are resistant to:

(a) Black tip **(b) Malformation**

(c) Spongy tissue (d) None of these

1086. Bio-control agent for woolly apple aphid in apple is:

(a) *Aphilenus mali* (b) Lady bird beetle

(c) *Trichagramma* (d) *Blastophaga psens*

1087. Bitter pit in apple is caused due to deficiency of

(a) Mangonese **(b) Calcium**

(c) Boron (d) Zinc

1088. Bitterness in citrus juice is due to:

(a) Acid (b) Limonin

(c) Glucosides **(d)** Acid:Sugar ratio

1089. Bitterness in peach is due to:

(a) Sugar **(b) Prunasin acid**

(c) Maleic acid (d) Hydrocynin

1090. Biurate content in urea is:

(a) 15% **(b) 2%**

(c) 4% (d) 5%

1091. Bizzare Orange, a graft chimera has originated as a result of:

(a) Sweet orange and Sour orange

(b) Sour orange and Pummeto

(c) Grapefruit and Sour orange

(d) Sour orange and Citron

1092. Black Heart and Flogmore Early are varieties of

(a) Cherry (b) Apple

(c) Almond (d) Pear

1093. Black pepper, clove and cardamom are categorised as:

(a) Spices (b) Condiments

(c) Both these (d) None of these

1094. Black tip in mango is a physiological disorder due to:

(a) Carbon monoxide (b) Excess Boron

(c) Deficiency of Calcium (d) Chloride toxicity

1095. Black tip of mango is also known as:

(a) Chimney disease (b) Soft nose of mango

(c) Witch's broom (d) Seed jelly of mango

1096. Black tip of mango is caused by:

(a) Virus (b) Fungus

(c) Bacteria **(d) None of these**

1097. Black tip of mango is due to:

(a) A bacterium (b) A fungus

(c) A virus (d) **Toxic gases of brick kilns**

1098. Blood red Malta is commercially cultivated in which one of the following states of India?

(a) UP (b) AP

(c) Punjab **(d) MH**

1098. Blood Red variety of which highly suited with which rootstock in Punjab:

(a) Rough lemon (b) Sour orange

(c) Cleopatra mandarin (d) *Poncirus trifoliate*

1099. Blood Red" "Pineapple"' "Hamlin "Jaffa and "Valencia late are sweet orange varieties important in:

(a) Southern India (b) Central India

(c) Northern India (d) Western India

1100. Bodles Altafort a tetraploid banana is a result of controlled crossing between:

(a) Gross Michel and Pisang Lilin

(b) Gross Michel and Ney Poovan

(c) Gross Michel and Kadali

(d) None of these

1101. Bordeaux mixture was discovered by PA Milardet of 'France in 1882 following his chance observation of a farmer's practice for protection against:

(a) ***Plasmopara viticola* on grapevine**

(b) *Podosphara leuchtricha* on apple

(c) *Venturia inaequalis* on apple

(d) *Uncinula necator* on grapevine

1102. Botanically pitanga cherry is called as:

(a) *Syzigium communi* (b) *S. uniflora*

(c) *S. densiflora* (d) None of these

1103. Botanically banana fruit is a

(a) Drupe **(b) Berry**

(c) Pome (d) Nut

1104. Botanically fruit type of jackfruit is:

(a) Sorosis (b) Drupe

(c) Syconus (d) Stone

1105. Botanically loquat is a which type of fruit:

(a) Nut **(b) Pome**

(c) Berry (d) Drupe

1106. Botanically strawberry is in:

(a) Berry (b) me fruit

(c) Aggregate fruit (d) Sorosis fruit

1107. Botanically the anola fruit is:

(a) Drupe **(b) Berry**

(c) Pome (d) Sorosis

1108. Botanically the apple fruit is:

(a) Pome (b) Berry

(c) Grape (d) Sorosis

1109. Botanically the avocado fruit is:

(a) **Berry** (b) Drupe

(c) Pome (d) Sorosis

1110. Botanically the ber fruit is:

(a) **Drupe** (b) Berry

(c) Pome (d) Sorosis

1111. Botanically the fruit of pineapple is:

(a) Drupe (b) Pome

(c) Berry (d) **Sorosis**

1112. Botanically the fruit of Pomegranate is:

(a) Drupte (b) Berry

(c) **Balausta** (d) Hesperidium

1113. Botanically the grapefruit is:

(a) Berry (b) Pome

(c) **Hesperidium** (d) Sorosis

1114. Botanically the guava fruit is:

(a) Drupte (b) Hesperidium

(c) Sorosis (d) **Berry**

1115. Botanically the lemon fruit is:

(a) Drupe (b) Pome

(c) Berry (d) **Hysperidium**

1116. Botanically the litchi fruit is:

(a) Drupe (b) Berry

(c) **Nut** (d) Sorosis

1117. Botanically the mature fruit of litchi is:

(a) Nut (b) Berry

(c) Drupe (d) **Aril**

1118. Botanically the papaya fruit is:

(a) Drupe **(b) Berry**

(c) Pome (d) Sorosis

1119. Botanically the pear fruit is:

(a) Drupe **(b) Pome**

(c) Berry (d) Sorosis

1120. Botanically what a cashewnut fruit is?

(a) Pome (b) Berry

(c) Drupe **(d) Nut**

1121. Botanically guava is a:

(a) Drupe (b) Pome

(c) Berry (d) Nut

1122. Botanically, quince is:

(a) Pome (b) Hesperidium

(c) Drupe (d) Sorosis

1123. Botanically, sapota fruit is a

(a) Berry (b) Pome

(c) Drupe (d) Aggregate

1124. Botanically, the fruit of litchi is a:

(a) Berry (b) Drupe

(c) Pome **(d) Nut**

1125. Botanically the Pear fruit is a:

(a) Berry **(b) Pome**

(c) Drupe (d) Multiple fruit

1126. Bower system of training in grapes is also called as:

(a) Head system (b) Kniffin system

(c) Arbour system (d) Single Stalk system

1127. Branch of science that deals with study of principles of propagation and production of fruits is called:

(a) Olericulture (b) Floriculture

(c) Pomology (d) Entomology

1128. Breeding sweet orange x Willow leaf mandarin produced one of the following:

(a) Tangering (b) Citron

(c) Kinnow (d) Pummelo

1129. Bromelin" alkaloid is found in:

(a) Ber (b) Custard apple

(c) Pineapple (d) Sapota

1130. Bronzing in guava is associated with the deficiency of:

(a) Zn (b) Mn

(c) Mg (d) Cu

1131. Bt gene is incorporated for:

(a) Insect resistant (b) Disease resistance

(c) Herbicide resistance (d) Stress tolerance

1132. Budded trees of oranges start bearing from fourth year, but full crop can be had from:

(a) 7th year (b) 10th year

(c) 9th year (d) 11th year

1133. Bridge grafting is practiced to repair damage:

(a) Root **(b) Shoot**

(c) Both (d) None of these

1134. Bullock's Heart is used as rootstock in:

(a) Custard apple (b) Fig

(c) Guava (d) Aonla

1135. Bunchy top disease of banana is mainly transmitted by:

(a) Aphid (b) Fungus

(c) Bacteri (d) Insects

1136. Bunchy top entered into India from:

(a) USA (b) Japan

(c) Australia **(d) Sri Lanka**

1137. Bunchy top is a disease of:

(a) Citrus (b) Guava

(c) Banana (d) Pineapple

1138. Bur knot formation is commonly found in:

(a) Apple (b) Pear

(c) Cherry (d) Almond

1139. Cashew fruit is called:

(a) Cashew apple (b) Cashew ball

(c) Cashew nut (d) Cashew lob

1140. Calcium carbide release:

(a) Benzene gas (b) Ethylene gas

(c) Methane gas (d) **Acetylene gas**

1141. California Advance, Large Round, Fire Ball, Thames Pride and Tanaka are partial self incompatible varieties of which of following?

(a) Loquat (b) Banana

(c) Citrus (d) Guava

1142. "California Advance" variety of loquat was introduced from:

(a) Brazil (b)' China

(c) USA **(d) Australia**

1143. Callus formation wounded plant parts and initiation of new growing points on apices of root and stem occur due to

(a) Miotic cell division (b) Meiotic cell division

(c) Both these (d) None of these

1144. Calyptra is a term associated with which of the following fruit crops?

(a) Litchi
(b) Grape
(c) Aonla
(d) Mango

1145. Calyx end rot is disorder caused by calcium deficiency and is observed in:

(a) Pecan
(b) Persimmon
(c) Apple
(d) Guava

1146. Canker in guava is caused by:

(a) Virus
(b) Fungi
(c) Pome
(d) Bacteria

1147. Canker is common in which one of the following:

(a) Citrus
(b) Mango
(c) Ber
(d) Aonla

1148. Canker is most serious problem of:

(a) Orange
(b) Lemon
(c) Lime
(d) Grapefruit

1149. Canning is a method of:

(a) Heat processing
(b) Non heat processing
(c) Sterilization
(d) Any other

1150. Canning variety of grape:

(a) Thompson Seedless
(b) Pusa Seedless
(c) Champion
(d) Beauty Seedless

1151. Caprification is found in:

(a) Phalsa
(b) Fig
(c) Sapota
(d) None of these

1152. Carambola (*Averrhoea carambola*) or Five-corner fruit belongs to family:

(a) Anacardiaceae
(b) Oxalidoceae
(c) Caricaceae
(d) Bombacacae

1153. Carambola is rich in which of the following organic acids?:

(a) Malic
(b) Oxalic
(c) Tartaric
(d) Citric

1154. Cardinal' grape is a cross between:

(a) Vinifera and laburusca
(b) Tokey and Ribier
(c) Anab-e- Sahi and Thompson Seedless
(d) Perlette and Himrod

1155. Carica papaya can successfully cross with F_1 which following species of papaya?: (1). *Carica cauliflora* (2) *C. monoica* (3). *C. candamarcensis* (4). *C. microcarpa*

(a) 1 and 2
(b) 2 and 3
(c) 1, 2 and 3
(d) All the above

1156. *Carpomyia vesuviana* damages extensively the following fruit

(a) Pomegranate
(b) *Ziziphus*
(c) Mandarin
(d) Guava

1157. *Carrisa grandiflora* (Natal plum) is native to:

(a) South Africa
(b) China
(c) India
(d) None of these

1158. Cashew has originated from:

(a) Brazil
(b) India
(c) Caspian Sea
(d) South East Asia

1159. Cavendish banana is propagated through:

(a) Seed
(b) Sucker
(c) Leaf cutting
(d) Stem cutting

1160. Cell division continues till maturity in fruit development of:

(a) Peach
(b) Strawberry
(c) Avocado
(d) Mango

1161. Central Institute for Sub Tropical Horticulture, Lucknow was initially

(a) Central Institute for Mango Research

(b) Central Mango Research Station

(c) Central Mango Research Institute

(d) Central Mailing Research Station

1162. Central Institute for subtropical Horticulture is located

(a) Putter (Karnataka)

(b) Lucknow, UP

(c) Nasik (Maharashtra)

(d) Hyderabad (Andhra Pradesh)

1163. Central Institute for Temperate Horticulture is located at:

(a) Solan **(b) Srinagar**

(c) Kullu (d) Chaubattia

1164. Central Plantation Crop Research Institute was established in:

(a) 1949 (b) 1970

(c) 1954 **(d) 1965**

1165. Central Potato Research institute was shifted to Shimla in:

(a) 1956 **(b) 1948**

(c) 1957 (d) 1960

1166. Central Tuber Crop Research Institute was established in:

(a) 1963 (b) 1964

(c) 1964 (d) 1996

1167. Centre of origin of Guava;

(a) Chile to Peru (b) China

(c) Africa (d) Paraguay

1168. Chakaiya is a variety of:

(a) Grape Fruit **(b) Amla**

(c) Lime (d) Sweet Orange

1169. Chakarpara and Charamgz are improved cultivar of:

(a) Peach (b) Apple

(c) Apricot (d) Pear

1170. Chakrotra is a cultivar of:

(a) Lime (b) Mandarin

(c) Grape Fruit (d) Sweet orange

1171. Chaubattia Princess and Chaubattia Anupam are hybrids of:

(a) Apple (b) Pear

(c) Licthi (d) Peach

1172. Chaubattia Agrim is an early maturing hybrid of which of which of the following:

(a) Peach **(b) Apple**

(c) Plum (d) Pear

1173. Chaubattia Alankar and Chaubattia Madhu are cultivars of:

(a) Apricot (b) Plum

(c) Peach (d) Apple

1174. Cheema Sahibi (Selection 7), a grape variety selected by Dr. G.S Cheema, which was the parent material of this variety?

(a) Thompson Seedless (b) Roa Sahibi

(c) Pusa Seedless (d) **Pandhari Sahibi**

1175. Cherry is propagated by whip grafting or tongue grafting on seedlings of:

(a) Wild apricot (b) Walnut

(c) Peach (d) **Wild cherry called paja**

1176. Chemical formula for potassium metabisulphite:

(a) KMS (b) $KS0_2$

(c) K_2S0_2 **(d) $K_2S_20_7$**

1177. Cherries are grown in Simla Hills, Kulu valley and Kashmir at elevations of higher than:

(a) 1,500 metres (b) 2,000 metres

(c) 2,100 metres (d) 5,200 metres

1178. Cherries are of two types, namely:

(a) Sweet and sour (b) Black and white

(c) Temperate and tropical (d) Red and green

1179. Chestnut is grown in which of the following fruit growing region?

(a) Temperate (b) Tropical

(c) Semi arid (d) Sub tropical

1180. Chicklet is obtained from the latex of:

(a) Sapota (b) Fig

(c) Papaya (d) Mango

1181. Chief centre of walnut growing state in India is:

(a) Punjab (b) HP

(c) J&K (d) UP

1182. Chief characteristic of arid zone is:

(a) Cloudless weather

(b) High humidity

(c) Low precipitation

(d) Intensive water loss from soil

1183. Chiku fruits can be stored for 2-3 weeks at a temperature of:

(a) 32-35°F (b) 42-45°F

(c) 52-55°F **(d) 67-70°F**

1184. Chiku is a Cross pollinated fruit crop because of:

(a) Dioecious (b) Protoandrous

(c) Protogynous (d) Heterostyly

1185. Chiku is commonly propagated by:

(a) Hard wood cutting **(b) Inarching**

(c) T-budding (d) Seeds

1186. Chiku is Cross-pollinated crop and pollination is mediated through:

(a) Insects (b) Birds

(c) Wind (d) Human beings

1187. Chiku is propagated by:

(a) Seed **(b) Vegetative propagation**

(c) Both a and b (d) None of these

1188. Chiku is propagated on root stock of:

(a) *Mangifera indica* (b) *Melia azadirach*

(c) *Jasmine sambac* ***(d) Manikara hexandra***

1189. Chinese guava is botanically:

(a) *Psidium guajava* ***(b) P. Friedrichsthalianum***

(c) *Weianum* (d) None of these

1190. Chok throat in banana is due to:

(a) Low temperature (b) Low humidity

(c) High humidity (d) None Of these

1191. Chromosome number (2n) of lemon is:

(a) 18 (b) 22

(c) 28 (d) 38

1192. Cidar is a product prepared from:

(a) Pear (b) Peach

(c) Apple (d) Almond

1193. CISH-MI an improved mango, is a cross between:

(a) Amrapali x Janaradhan Pasand

(b) Amrapali x Sensation

(c) Dashehari x Sensation

(d) Dashehari x Janardhan pasand

1194. Citranges, the potential rootstock of citrus has originated as a result of cross between:

(a) *Citrus sinensis* and Trifoliate orange

(b) Sweet orange and Mandarin

(c) Trifoliate orange and Sweet orange

(d) Trifoliate orange and Mandarin

1195. Citrus canker is a serious disease caused by:

(a) Mycoplasma **(b) Bacterium**

(c) Fungus (d) Virus

1196. Citrus die-back is mainly caused by:

(a) Virus (b) Water stress

(c) Fungi and other factors (d) Bacteria

1197. Citrus fruit can be preserved for a few weeks at room temperature by:

(a) Submerging them is cold water and then drying

(b) Keeping them in well ventilated room

(c) Putting them in an atmosphere rich in CO_2

(d) By wrapping individual fruit in HDP envelopes

1198. *Citrus limonia* is commonly known as:

(a) Rangpur lime (b) Acid lime

(c) Persian lime (d) Rough Lemon

1199. *Citrus myrtifolia* is originated as a result of bud mutation from which species of citrus:

(a) *Citrus reticulata* (b) *C. sinensis*

(c) *C. aurantium* (d) *C. nobilis*

1200. Citrus originated in:

(a) South Africa **(b) South-East China**

(c) West Indies (d) America

1201. Citrus species indigenous to India:

(a) ***Citrus macroptera*** (b) C. *sinensis*

(c) C. *grandis* (d) C. *reticulata*

1202. Citrus species relative resistant to sunburn and sun scold is:

(a) Kinnow (b) Lime

(c) Sweet orange (d) Lemon

1203. Citrus species which originated in New World:

(a) C. *aurantium* (b) C. *unshiu*

(c) C. *aurantifolia* **(d) C. *paradise***

1204. Clementine tongarine is:

(a) Self-incompatible (b) Self-compatible

(c) Parthenogenesis (d) Seedless

1205. Clonal rootstocks of apple can be propagated through:

(a) Cutting **(b) Stooling**

(c) Air layering (d) Tongue grafting

1206. Close planting of banana increases the incidence of:

(a) Bacterial wilt (b) Bunchy top

(c) Sigatoka leaf spot **(d) Finger tip**

1207. CO-1 banana hybrid is similar to which of the following?

(a) Lal kela (b) Poovan

(c) Dwarf Cavendish (d) Virupakshi

1208. CO1, a hybrid of banana is the cross between, laden (AAB) Kadali (AA) and *M. balbisiana* (BB) Manthan, the correct sequence of multivarietal cross is:

(a) F1 from cross between Kadali and Laden was crossed with *M. balbisiana.*

(b) F1 from the cross between Laden and *M. balbisiana* as crossed with Kadali

(c) *M. balbisiana* and Kadali is crossed with Laden

(d) None of the above

1209. Cock's comb is a physiological disorder associated with:

(a) **Sapota** (b) Litchi
(c) Pear (d) Apple

1210. Cocoa belongs to family:

(a) Bombaceae (b) **Sterculiaceae**
(c) Oxoilidaceae (d) Proleaceae

1211. Cocoa is originated is South America. It was introduced in India by:

(a) British's East India Company (b) Muslim Invaders
(c) Portuguese (d) None of the above

1212. Coconut Development Board belongs to:

(a) DARE (b) ICAR
(c) Ministry of Agriculture **(d) Ministry of Commerce**

1213. Coconut Development Board is located at:

(a) Patna (b) Kota
(c) Cochin (d) Hyderabad

1214. Coconut attains full maturity within months from fruit set

(a) 6-8 (b) 10-12
(c) > 16 **(d) 13-15**

1215. Coffee plants are matured at the age of how many years:

(a) 15 (b) **10**
(c) 20 (d) 4

1216. Coffee rust is caused by:

(a) *Uromyces* sp. (b) ***Hemeleia vasatrix***
(c) *Coronartium ribicola* (d) *Ravaenelia embelica*

1217. California Advance, Large Round, Fire Ball, Thomas Pride and Tanaka are partial self-incompatible varieties of.

(a) Loquat (b) Banana
(c) Citrus (d) Guava

1218. Colour pigment in the flavour of mature ripe orange fruit is:

(a) Hesperidine (b) Naringin

(c) Carotene (d) Chlorophyll

1219. Colt is a rootstock of:

(a) Cherry (b) Pear

(c) Plum (d) Apple

1220. Commercial grape cultivar of Karnataka is:

(a) Beauty Seedless (b) Perlette

(c) Thompson **(d) Bangalore Blue**

1221. Commercial method of pear propagator is:

(a) Hardwood cuttings (b) Layering

(c) Tongue grafting (d) Chip budding

1222. Commercial method of vegetative propagation in walnut is

(a) T-budding **(b) Patch budding**

(c) Air Layering (d) Stooling

1223. Commercial propagation of papaya is done through:

(a) Budding (b) Layering

(c) Seeds (d) Cutting

1224. Commercially cultivated ber species in India is:

(a) *Z. rotundifolia* (b) *Z. oxylopyrus*

(c) *Z. Vulgaris* **(d) *Z. mauritiana***

1225. Commercially peach is propagated by:

(a) Patch budding (b) Inarching

(c) Air layering (d) Stooling

1226. Commercially pomegranate is propagated through:

(a) Hard wood cutting (b) Grafting

(c) Budding (d) Seeds

1227. Common deficient micronutrient noticed in litchi is:

(a) Iron (b) Copper

(c) Zinc (d) Manganese

1228. Common disorder of Sapota is:

(a) Fruit cracking **(b) Cock's comb**

(c) Button shedding (d) Water core

1229. Common rootstock for budding of mandarin is:

(a) *C. jambhiri* (b) *C. sinensis*

(c) *C. limonia* (d) *C. aurantium*

1230. Commonly practiced standard method of curing harvested banana fruits is:

(a) Sun-curing (b) Straw-curing

(c) Smoke curing (d) Fungal curing"

1231. Compact Lambert, a sweet cherry variety is:

(a) Self-fruitful (b) Self-unfruitful

(c) Self-fertile (d) Self-sterile

1232. Concord, Isabella and Rogers varieties of grape belong to species:

(a) *Vitis vinifera* (b) *Muscadinia rotundifolia*

(c) *V. labrusca* (d) *V. rupestris*

1233. Contour system of orchard planting is generally followed in:

(a) Punjab **(b) Hills**

(c) Saline soils (d) UP

1234. Controlled crossing between which of the following resulted in 'Boodles Altafort' a tetraploid banana?

(a) Gross Michel and Pisang Lilin

(b) Gross Michel and Ney Poovan

(c) Gross Michel and Kadali

(d) None of the above

1235. Cork spot disorder in pear is caused due to deficiency of:

(a) Boron (b) Manganese

(c) Calcium (d) Iron

1236. Creeper is a dwarf variety of which fruit?

(a) Apple **(b) Mango**

(c) Grape (d) Peach

1237. Cricket Ball is variety of:

(a) Litchi (b) Mangosteen

(c) Sapota (d) Durion

1238. Cricket Ball, Guthi, Kallipati, Oval, Baramasi are the variety of:

(a) Pummelo (b) Plum

(c) Sapota (d) Tamarind

1239. "Cricket Ball 'Dwaropudi', 'Bangaloria', 'Vavilavalasa', 'Jonnavalas 'Kirti-barati', 'Pot, 'Kalipatti 'Chatri' are the varieties of:

(a) Sapota (b) Muskmelon

(c) Papaya (d) Mango

1240. Criteria for the essentiality of nutrient for plants was given by:

(a) Arnon (b) Wilcox

(c) Liebig (d) None

1241. Crop regulation (Bahar Treatment) in citrus, pomegranate and guava is achieved by:

(a) Withholding of water

(b) Root and shoot pruning

(c) Deblossomihg by application of NAA

(d) All the Above

1242. Cross between which of the following resulted in 'Stella' a self fertile cultivar of cherry?

(a) Lamberi & John Inns Seedling 2420

(b) Sue x Stella

(c) Lambert and Colt

(d) None of the above

1243. Cross between which of the following resulted in Kinnow?

(a) *Citrus nobilis* and *C. deliciosa*

(b) *C. nobilis* and C. *Reticulata*

(c) *C. reticulata* and *C. sinensis*

(d) *C. Miiobilis* and *C. sinensis*

1244. Cross between which of following results in plumeot?

(a) Apricot and plum

(b) Plum and apricot

(c) Peach and Apricot

(d) Peach and almond

1245. 'Cross pollination in loquat is favoured by self incompatibility, whereas in sapota, it is due to:

(a) Protogyny

(b) Homogamy

(c) Protandry

(d) None of these

1246. Cross pollination which of the following resulted in IC2 and S19, triploid banana hybrids?

(a) Gross Michel and Pisang Lilin

(b) Gross Michel and M. acuminate

(c) Gross Michel and Kadali

(d) None of the above

1247. Crosses for the edible strawberry were made by:

(a) Thomas Knight Andrew

(b) Simmands and Shepherd

(c) Bompard and Casterman

(d) None of the above

1248. Crossing between which of following resulted in Ratna variety of the mango?

(a) Alphonso x Banganpalli

(b) Neelum x Langra

(c) Dashehari x Neelum

(d) Neelum x Alphonso

1249. Crossing between which of the following edible strawberry has originated as a result of a cross between

(a) *F. chiloensis* and *F. vesca*

(b) *F. chiloensis* and *Fragaria virginiana*

(c) *F. virginiana* and *F. vescea*

(d) None of the above

1250. Crossing between which of the following resulted in 'Keiffer' and 'Le Conte' pears?

(a) ***Pyrus communis* and *P. pyrifolia***

(b) *P. serotina* and *P. communis*

(c) *P. pyrifolia* and *P. pashia*

(d) None of the above

1251. Crossing between which of the following resulted in Arka Sahan?

(a) ***Annona atemoya* x *A. squomosa***

(b) *A. reticulata* x *A. chrimoya*

(c) *A. reticulata* x *A. glabru*

(d) *A. reticulata* x *A. squamosa*

1252. Crossing between which of the following resulted in development of 'Santa Rosa' a popular plum variety?

(a) ***Prunus simonii, P. salicina* and *P. americana***

(b) P. *salicina*. P. *domestica* and P. *americana*

(c) *P. cerasifera, P salicina* and *P. domestica*

(d) None of the above

1253. 'Crossing between which of the following resulted in super dwarfing rootstock of apple, M-27

(a) **M9 x M13** (b) M12 x M16

(c) M-12xM-15 (d) M-20 x M-9

1254. Crossing between which of the following resulted in the evolution of 'Manjeera' a mango hybrid?

(a) **Rumani and Neelum** (b) Neelum and Dashehari

(c) Rumani and Mulgoa (d) Dashehari and Neelum

1255. Crossing between which resulted in variety 'Arka Aruna' of mango?

(a) **Banganpalli x Alphonso**

(b) Alphonso x Janardhan Pasand

(c) Alphonso x Banganpalli

(d) Alphonso x Dashehari

1256. Crossing over during meiosis results in

(a) Help mutation (b) Promoting

(c) Breaking linkage (d) None

1257. Crown grafting and side grafting are the methods of:

(a) Improving mango varieties

(b) Producing protoplasmic hybrids

(c) Propagating grape vines

(d) Propagating Dahlia

1258. Cryopreservation is done in liquid nitrogen at a temperature of:

(a) 0°C (b) 4°C

(c) 200°C **(d) -196°C**

1259. Cultivated strawberry is a:

(a) Tetraploid (b) Hexaploid

(c) Octaploid (d) Tetraploid

1260. Custard apple a multiple fruit the botanical form' of this fruit is known as:

(a) Etaerio of achenes **(b) Etaerio of drupelets**

(c) Pome (d) Etaerio of berries

1261. Cutting is a method of:

(a) Plant propagation

(b) Slicing fruits for processing

(c) Both these

(d) None of these

1262. Cutting, grafting, layering, separation and division are methods of:

(a) Vegetative propagation (b) Sexual propagation

(c) Both these (d) None dipese

1263. Cytokines cause:

(a) Cell division (b) Growth retardation

(c) Rooting (d) Cell abortion

1264. Dakh" is a popular grape variety of:

(a) Punjab (b) AP

(c) TN (d) KN

1265. Dashehari x Neelam are the parents of which mango variety?

(a) Arka Aruna (b) Mallika

(c) Amarpali (d) Sindhu

1266. Dashehari-51 differs from Dashehari because it is:

(a) Regular bearer

(b) Resistant to powdery mildew

(c) Tolerant to mango hopper

(d) Anthracnose tolerant

1267. Date palm is native to:

(a) Iran **(b) Iraq**

(c) Israel (d) Russia

1268. Date palm is propagated by:

(a) Off shoots (b) Suckers

(c) Both these (d) Budding

1269. Date-palm, phalsa, custard apple, bullock's heart, fig, Aonla and Jujube are the fruits of:

(a) Arid and semi-arid area (b) Temperate areas

(c) Humid areas (d) Wet areas

1270. Deficiency of vitamin B leads to:

(a) Scurvy **(b) Beri-Beri**

(c) Pelagra (d) Night blindness

1271. Deficiency of vitamin "C" (ascorbic acid) in human diet leads to the development of:

(a) **Scurvy** (b) Beri-beri

(c) Pellagra (d) Goitrel 18

1272. Deficiency of which nutrient results in 'Yellow Spot' disorder in Citrus?

(a) **Mo** (b) S

(c) B (d) Ca

1273. Deficiency of which nutrient cause's bitter pit in apple?

(a) Manganese (b) **Calcium**

(c) Boron (d) Zinc

1274. Deficiency of which nutrient element causes 'Little leaf of mango?

(a) **Zn** (b) K

(c) Cu (d) Fe

1275. Deficiency of which nutrient element results in 'Smelling' 'Rosette'formation and 'internal chlorosis' in young pear leaves:

(a) Fe (b) **Zn**

(c) Mn (d) Cu

1276. Deficiency of which nutrient element results in Litchi leaf scorching?

(a) N (b) P

(c) Ca (d) **K**

1277. Deficiency of which nutrient lead to Bronzing in guava?

(a) **Zn** (b) Mn

(c) Mg (d) Cu

1278. Deficiency of which of the following nutrient is the cause of interveinal chlorosis in fruit crops?

(a) Nitrogen (b) **Magnesium**

(c) Potassium (d) Calcium

1279. Deficiency of which of the following vitamins causes scurvy?

(a) Vit A (b) Vit B

(c) Vit C (d) Vit D

1280. Deficiency of which plant nutrient causes Millerandage' a disorder of grape?

(a) B (b) Ca

(c) Fe (d) Zn

1281. Deficiency or toxicity of which nutrient element cause leaf scorching in Mango?

(a) N deficiency **(b) K deficiency**

(c) Chloride Deficiency (d) Chloride Toxicity

1282. Deficiency symptom of which of the following elements appears on young leaves?

(a) S (b) Mo

(c) N (d) Zn

1283. Deficiency symptoms of calcium on plants are first noticed at:

(a) Basal leaves **(b) Middle leaves**

(c) Terminal leaves (d) All leaves

1284. Deficiency of which one of the following has a dwarfing effect on vegetative growth: also does not allow seed formation:

(a) Calcium (b) Copper

(c) Iodine **(d) Zinc**

1285. Defoliation in Annona results in good crop. It is achieved by the spray of 0.05% (w/v):

(a) K_2S0_4 **(b) KI**

(c) KIO3 (d) KCI

1286. Degreening in citrus is done with the application of which of the following?

(a) GA_3 (b) Auxins

(c) Ethephon (d) Cytokines

1287. Delight Gold is a variety of:

(a) Guava (b) Papaya

(c) Grape (d) Mango

1288. Desi emperor, Nagpur orange Khasi orange, Coorg orange and Sikkim orange are varieties of:

(a) Mandarin orange **(b) Sweet orange**

(c) Grape fruit (d) Orange lime

1289. Dhola Diwani; variety of chiku is common in:

(a) TN (b) WB

(c) UP **(d) MH**

1290. Dholka and Kandhari are the varieties of:

(a) Loquat (b) Amla

(c) Pomegranate (d) Jamun

1291. Die back of citrus can be controlled with the application of:

(a) Bordeaux mixture (b) Endosulfan

(c) Zinc sulphate (d) BHC

1292. Deficiency of which leads to the symptoms: (i) some species are not able to absorb nitrates (ii) generally immobile from one part of the plant to another:

(a) Calcium (b) Iron

(c) Manganese (d) Magnesium

1293. Digraset grape rootstock is tolerant to which stress?

(a) Salt (b) Pierce Disease

(c) Nematode **(d) Phylloxera**

1294. *Diospyrus lotus* is used as rootstock in

(a) Walnut **(b) Persimmon**

(c) Pecan nut (d) Avocado

1295. Diploid cultivars of apple are generally:

(a) **Self-fruitful** (b) Self-unfruitful

(c) Self-sterile (d) Cross-fertile

1296. D-leaf is a best indicator of nutrient status of:

(a) Banana (b) **Pineapple**

(c) Mango (d) Citrus

1297. DNA contains following number of nitrogenous bases repeated in various sequences:

(a) One (b) Two

(c) **Four** (d) Ten

1298. Domestic quarantine exist in India for:

(a) **Potato eelworm** (b) Panama wilt of banana

(c) Sanjose scale (d) Sugarbeet eelworm

1299. Dormex/hydrogen cynamide is used for early bud break in:

(a) **Grape** (b) Mango

(c) Citrus (d) Guava

1300. Double sigmoid growth curve is found in:

(a) Apple (b) Pear

(c) Ber (d) **Fig**

1301. Double working is commonly followed in:

(a) Plum on its own roots

(b) **Pear on quince rootstock**

(c) Mango on Vellaikullamban

(d) Apple on crab apple rootstock

1302. Double working is commonly followed on:

(a) Peach (b) Apple

(c) **Mango** (d) Grape

1303. Dried litchi called litchi nut are commercially produced in:

(a) India **(b) China**

(c) Brazil (d) Indonesia

1304. Drought resistant rootstocks of grapes?

(a) Dog ridge; 1613 (b) St George & Salt Creek

(c) Salt Creek and Dogridge (d) St George and Digraset

1305. Dual factor stimulus of flowering hypothesis/dual factor hypothesis:

(a) Simmons Shepherd **(b) Chakraborty**

(c) Gross Michel (d) None

1306. Durian (Civet fruit), *Durian zibenthinus* belongs to which family:

(a) Bomabaceae (b) Sapindaceae

(c) Guttiferae (d) Ebenaceae

1307. Durian is originated:

(a) Indo-Burma (b) **Malayan Archipelago**

(c) Chinese (d) Asia Minor

1308. During vegetative phase of plant growth the cell division is:

(a) Meiotic (b) **Mitotic**

(c) Both these (d) None of these

1309. During which year Kinnow was introduction in India?

(a) 1940 **(b 1959**

(c) 1970 (d) 1985

1310. During which year was Pineapple introduced in India?

(a) 1248 **(b) 1348**

(c) 1448 (d) 1548

1311. Dashehari and Goapalbhog are the mango varieties popular in:

(a) North India (b) Karnataka

(c) Assam (d) Gujarat

1312. Dwarf Cavendish is a variety of:

(a) Tomato (b) Mango

(c) Banana (d) Grape

1313. Dwarf mango cultivar is:

(a) Chausa (b) Neelam

(c) Leangra **(d) Amrapali**

1314. Dwarf mango suitable for kitchen garden is:

(a) Alphonso (b) Chausa

(c) Dashehari **(d) Amarpali**

1315. Dwarf mutant variety of papaya is:

(a) CO-1 (b) Pusa dwarf

(c) Pusa Nanha (d) CO-6

1316. Dwarf rootstock of ber is:

(a) *Z. ratundifolia* (b) *Z. rugose*

(c) *Z. xylopyrus* (d) None of the above

1317. Dwarfing root stock in mango is:

(a) Olour **(b) Totapuri Red Small**

(c) Alphonso (d) Dashehari

1318. Dwarfing rootstock of apple is:

(a) M-27 (b) M-13

(c) M-21 **(d) M-9**

1319. Dwarfness in papaya is a desirable character. What is correct about its inheritance?

(a) It is recessive (b) Dominant

(c) Epistotic (d) None of the above

1320. Dwarfness in papaya is controlled by:

(a) Recessive gene (b) Co-dominant gene

(c) Dominant gene (d) Duplicate gene

1321. Dwarfness is a desirable character for selection in fruit crop. The pre-selection criterion for dwarfness: (1) Leaf size (2) Respiration rate (3) Phloem/Xylem ratio (4) Stomatal density on abacial leaf surface

(a) 1 and 2 only (b) 3 only

(c) 2 only **(d) 3 and 4**

1322. Earliest ripening commercial grape variety:

(a) Pearl of Casaba (b) Perlette

(c) Charat Early (d) Beauty Seedless

1323. Early post-blossom fruit drop in apple is known as:

(a) May drop (b) June drop

(c) July drop (d) August drop

1324. Early ripening dwarf yellow sweet, small fruited variety of pineapple good for table purpose is:

(a) Abakka (b) Sugar leaf

(c) Mauritius **(d) Queen**

1325. Early ripening grape variety suited for cultivation in North India is:

(a) Gulabi (b) Bhokir

(c) Perlette (d) Anab-e-Shahl

1326. Early ripening pomegranate variety is:

(a) Dholka **(b) Jalore seedless**

(c) Ganesh (d) Alandi

1327. Early Shipley is a variety of:

(a) Apple **(b) Apricot**

(c) Almond (d) Avocado

1328. Edible banana fruit is seedless due to:

(a) Embryo abortion (b) Absence of ovule

(c) Vegetative parthenocarpy (d) Stimulated parthenocarpy

1329. Edible part of rambutan:

(a) **Aril** (b) Seed
(c) Endocarp (d) Mesocarp

1330. Edible portion in litchi:

(a) Mesocarp (b) **Fleshy Aril**
(c) Endocarp (d) Seed

1331. Edible portion of pineapple:

(a) Thalamus (b) Peduncle
(c) **Bracts and Perianth** (d) Drupelets

1332. Embryo phase of a plant begins with:

(a) Formation of zygote (b) Formation embryo
(c) Formation of seeds (d) **Pollination**

1333. Embryo rescue technique can be used for production of:

(a) Seedless apple (b) **Seedless litchi**
(c) Seedless grapes (d) Seedless pomegranate

1334. End of the monsoon is the best time for mango plantation In:

(a) **Heavy rainfall areas** (b) Light rainfall areas
(c) Arid areas (d) Semi arid areas

1335. Endoxerosis is due to:

(a) High temperature (b) Excess water
(c) **Water deficiency** (d) Low temperature

1336. 'Endoxerosis' a disorder associated with water deficiency. It is associated with which of the following fruit crop?

(a) Apple (b) **Lemon**
(c) Pear (d) Guava

1337. Enzyme present in dried latex of papaya is:

(a) Xanthain (b) Protease
(c) Anthocyanin (d) **Papain**

1338. *Eriobotrya japonica* is the botanical name of:

(a) Sapota **(b) Loquat**

(c) Wood apple (d) lasora

1339. Essential ingredients for making jelly are:

(a) Lemon fruit, sugar, water, sodium benzoate

(b) Fully ripe guava, sugar, acid water

(c) Pectin, sugar, acid, water

(d) Immature plum fruit, water KMS

1340. Evergreen fruit like mango, guava and jack fruit should be planted in the month of:

(a) February-March (b) June-July

(c) May (d) December

1341. Exanthena disorder in citrus is due to:

(a) Copper deficiency (b) Zinc deficiency

(c) Calcium deficiency (d) Molybdenum deficiency

1342. Export variety of apple in India is:

(a) Netwon Wonder **(b) Golden Delicious**

(c) Golden Russet (d) Rome beauty

1343. Export variety of mango is:

(a) Langra (b) Baneshan

(c) Alphanso (d) Bombay Green

1344. False bloom, is a disorder associated with excessive water. It is observed in which of the following fruit crops:

(a) Strawberry **(b) Craneberry**

(c) None of these (d) Raspberry

1345. Family of Guava:

(a) Rosaceae (b) Caricaceae

(c) Myrtaceae (d) Euphorbiaceae

1346. Family of Pineapple:

(a) **Bromiliaceae** (b) Anacardiaceae
(c) Juglandaceae (d) Ebenaceae

1347. Fasciation/flatted fruit, multiple crown and sun scald are the disorders of:

(a) **Pineapple** (b) Grape
(c) Guava (d) Mango

1348. Fastening of vines to a upright support of a tree called panagara (*Erithrina indica*) is known as:

(a) Kniffen system (b) Head system
(c) **Single stake system** (d) Cordon system

1349. Female gametes of plants are called:

(a) Ova (b) **Ovaries**
(c) Pollen grains (d) Anthers

1350. Femaleness is papaya is controlled by genotype:

(a) M_1M (b) **M_2M**
(c) MM_2 (d) mm

1351. Fertilization occurs in:

(a) Anthers (b) **Flowers**
(c) Ovaries (d) Stamens

1352. Filbert can be propagated through:

(a) Tip layering (b) **Compound layering**
(c) Stool layering (d) Simple layering

1353. Ficin (Rennin) has 30-100 times milk clotting activity than animal rennin. This is present in the latex of which fruit crop?

(a) **Fig** (b) Pear
(c) Apple (d) Peach

1354. Fig in India is grown in:

(a) MP (b) Punjab

(c) Kerala **(d) MH**

1355. First President of Horticultural Society of India:

(a) Dr. G. S. Randhawa (b) Dr. G. S. Cheema

(c) Dr. S. K. Mukherjee (d) None of these

1356. Five fruit corer /Star fruit is botanically :

(a) Averrhoa carambola (b) *Averrhoa acidus*

(c) *Artocarpus bilimbi* (d) *Artocarpus lakoocha*

1357. False bloom, is a disorder associated with excessive water. It is observed in which of the following fruit crops:

(a) Strawberry **(b) Cranberry**

(c) None of these (d) Raspberry

1358. Fleshy pericarp is an edible portion of:

(a) Fig (b) Papaya

(c) Custard apple (d) Datepalm

1359. Flordasun is the variety of:

(a) Apple (b) Plum

(c) Apricot **(d) Peach**

1360. Flower bud differentiation in mango in north India:

(a) Oct-Nov. (b) Sep-Oct

(c) Jan-Feb (d) March-April

1361. Flower thinning in grapes is done with:

(a) Ethylene (b) Sevin

(c) NAA **(d) GA_3**

1362. Flowering in mango can be suppressed by the application of:

(a) Auxin **(b) Gibberellins**

(c) Kinetin (d) Paclobutrazal

1363. Flying dragon is the most dwarfing rootstock for:

(a) **Citrus** (b) Apple

(c) Mango (d) Pear

1364. Following combination is true chiller:

(a) Apple, Pear (b) Peach, Plum

(c) **Cherry, Walnut** (d) All

1365. Following is the dwarfing rootstock of apple:

(a) M25 (b) M 15

(c) **M9** (d) M1

1366. For arid climatic conditions which fruit should not be selected?

(a) Date palm (b) Ber

(c) Pomegranate (d) **Banana**

1367. For avoiding floral malformation and taking normal fruit crop mango trees are sprayed with:

(a) Zineb (b) **Gibrelic acid**

(c) Karathane (d) CCC

1368. For complete preservation the percentage of sugar in fruit jam should be nearly

(a) 90% (b) **70%**

(c) 50% (d) 30%

1369. For developing resistance against which insect pest Bt- gene is used?

(a) Lepidoptera insects (b) **Coleoptera insect**

(c) Homenoptera insect (d) None of these

1370. For drying apricot fruits are:

(a) Picked raw

(b) Picked half ripe

(c) **Picked when fully ripe**

(d) Allowed to drop on ground

1371. For hand pollination in date-palm 10 male trees are sufficient for female plants:

(a) 50 (b) 75

(c) 100 (d) 125

1372. For jelly making the best fruit is:

(a) Sour apple (b) Ripe banana

(c) Ripe mango (d) Ber

1373. For jelly preparation very important fruit is:

(a) Guava (b) Mango

(c) Apple (d) Grapes

1374. For malta blood red, the best root stock is:

(a) *Citrus jambheri* (b) Karna khatta

(c) Rangpur lime (d) Cleoptera mandarin

1375. For pickle making one can use maximum:

(a) 40% salt (b) 30% salt

(c) 25% salt **(d) 35% salt**

1376. For production of which of the seedless plants Embryo rescue technique can be used?

(a) Seedless apples (b) Seedless litchi

(c) Seedless grapes **(d) all the above**

1377. For propagation of plants cutting may be taken from:

(a) Stems, roots (b) Eaves

(c) Leaf buds **(d) All these**

1378. For raising new plants by vegetative reproduction parts of seedling plants are separated at:

(a) Juvenile phase (b) Adult phase

(c) Transition phase **(d) Any of these phases**

1379. For raising nursery for one acre of papaya plantation, quantity of seed required is about:

(a) 100-200 g (b) 50-100 g

(c) 150-200 g (d) 300 g

1380. For retention of seed viability in papaya the seed is stored at

(a) 2°C (b) 5°C

(c) 7°C **(d) 10°C**

1381. For successful grafting there should be:

(a) Scion-stock compatibility

(b) Intimate contact between scion-stock cambium regions

(c) Both these

(d) None of these

1382. For transport papaya and pineapple fruits are packed in single layers in bamboo baskets after

(a) Wrapping in straw (b) Wrapping in moist cotton

(c) Embedding in sand (d) Removing stalks

1383. For which of the following fruit trees, ideal cultivation conditions are indicated by the maximum 'head in fire and feet in water'?

(a) Aonla **(b) Date palm**

(c) Ber or Bor (d) Phalsa

1384. For which of the following fruits "veneer grafting" has been found to be the best method of propagation?

(a) Lime (b) Guava

(c) Mango (d) Banana

1385. For which of the following pests' tree bunding is effectively used in mango?

(a) Fruit fly (b) Fruit sucking moth

(c) Mealy bug (d) Aphid

1386. For which of the fruit crop flying dragon is the most dwarfing rootstock?

(a) **Citrus** (b) Apple

(c) Mango (d) Pear

1387. For which purpose Hydrogen cynamide (HCN) is used in grape?

(a) Berry thinning (b) Berry elongation

(c) **Bud break** (d) Induce dormancy

1388. Foster and Ruby grapefruit varieties have originated as:

(a) Bud Sport (b) Chance Seedling

(c) **Hybridization** (d) Nuclear Seedling

1389. *Fragaria vesca* is a wild strawberry of India. The ploidy level of this strawberry is:

(a) **Diploid** (b) Octaploid

(c) Tetroploid (d) Hexaploid

1390. First successful transgenic fruit plants are produced in:

(a) Pear (b) Cherry

(c) **Walnut** (d) Apple

1391. From which country 'Satgudhi' a variety of Sweet Orange is introduced?

(a) Indonesia (b) USA

(c) Malaysia (d) **India**

1392. From which country bunchy top of the banana entered into India?

(a) USA (b) Japan

(c) Australia (d) **Sri Lanka**

1393. From which country Wooly apple aphid, a serious pest of apple got its entry into India?

(a) USA (b) **England**

(c) China (d) Japan

1394. From which of the following guava berry develops?
   (a) Superior ovary
   (b) **Inferior ovary**
   (c) Intermediate ovary
   (d) a and b

1395. From which one of the following Cidar is prepared from?
   (a) Apple
   (b) Pear
   (c) Grape
   (d) Cashewnut

1396. From which place did Karonda originated?
   (a) India
   (b) South Africa
   (c) Sri Lanka
   (d) America

1397. Frontier is midseason variety of:
   (a) Apricot
   (b) Grape
   (c) Plum
   (d) Peach

1398. Fruit bud differentiation in apple in India occurs in:
   (a) Early to mid May
   (b) October-November
   (c) June-July
   (d) July-August

1399. Fruit candy is:
   (a) Agra-ka-petha
   (b) Aonla ka murabba
   (c) Murabba
   (d) Sugar candy

1400. Fruit cracking in litchi is due to:
   (a) Diseases
   (b) Nutrient deficiency
   (c) Heavy rains after long dry spell
   (d) Insect attack

1401. Fruit cracking in pear is due to deficiency of:
   (a) Boron
   (b) Manganese
   (c) Calcium
   (d) Iron

1402. Fruit cracking is a serious problem of:

(a) Litchi (b) Walnut

(c) Banana (d) Apple

1403. Fruit crop commercially micro propagated in India:

(a) Mango (b) Guava

(c) Wood apple (d) Grape

1404. Fruit crop grown in alkaline soils with minimum damage:

(a) Ber (b) Aonla

(c) Wood apple (d) Grape

1405. Fruit crop most suitable for soil with high pH

(a) Ber (b) Banana

(c) Guava (d) Grape

1406. Fruit crop resistant to wet soils:

(a) Peach (b) Apricot

(c) Quince **(d) Pear**

1407. Fruit crop suitable for cold desert of cold arid zone:

(a) Apricot (b) Apple

(c) Peach (d) Plum

1408. Fruit crop suitable for high soil pH:

(a) Aonla (b) Ber

(c) Both a & b (d) Pomegranate

1409. Fruit crop suited for arid regions is:

(a) Sweet orange **(b) Ber**

(c) Papaya (d) Mongo

1410. Fruit crops originated from Chinese centre of origin are:

(a) Peach, loquat, persimmon and litchi

(b) Papaya, avocado, sapota and banana

(c) Cherry, filbert pomegranate and fig

(d) Apple, pear strawberry and pecan

1411. Fruit crops originated from South American centre of origin are:

(a) Papaya, avocado, pecan and banana

(b) Guava, cashew and pineapple

(c) Peach, plum, quince and persimmon

(d) Mango, aonla, ber, phalsa

1412. Fruit drop in grape *Citrus sinensis* can be controlled by

(a) NAA (b) IBA

(c) 2,4,5-T **(d) 2,4-D**

1413. Fruit drop in mango can be controlled with the spray of:

(a) 2,4-D (b) BHC

(c) NaCI (d) GA_3

1414. Fruit growth follows a triple sigmoid curve in:

(a) Chinese gooseberry (b) Aonla

(c) Orange (d) Mango

1415. Fruit Introduced from Brazil to India.

(a) Guava **(b) Pineapple**

(c) Banana (d) None

1416. Fruit of *'Aegle marmelos'* is botanically:

(a) Hard-shelled berry (b) Drupe

(c) Single-seeded nit (d) Stone

1417. Fruit of cherry (*Prunus avium*) is botanically a

(a) Drupes (b) Berry

(c) Sorosis (d) Pome

1418. Fruit of fig is botanically 'Syconium', which is also a form of inflorescence. The individual fruitlet of fig is:

(a) Berry **(b) Achenes (Druplets)**

(c) Pome (d) None of these

1419. Fruit of karonda (*Carrisa carandus*) is berry. It belongs to family:

(a) Rutaceae
(b) Apocynaceae
(c) Rosaceae
(d) Euphorbiaceae

1420. Fruit on the basis of respiration can be classified as:

(a) Climacteric and non-climacteric fruits
(b) Non-climacteric fruits only
(c) Parthenocarpic fruits
(d) None of these

1421. Fruit set in apple is:

(a) 1%
(b) 5%
(c) 5-20%
(d) Less than 5%

1422. Fruit squash must have minimum of:

(a) 35% Juice
(b) 25% Juice
(c) 20% juice
(d) 45% juice

1423. Fruit stored in a cold chamber exhibit longer storage life because:

(a) Exposure to sun light is prevented
(b) CO_2 concentration in the environment
(c) The rate of respiration is decreased
(d) There is an increase in humidity

1424. Fruit thinning growth regulator:

(a) NAA
(b) IAA
(c) GA_3
(d) CK

1425. Fruits and vegetables are generally rich in dietary fibre which protects from:

(a) Fat deficiency
(b) Protein deficiency
(c) Colon cancer
(d) Carbohydrate deficiency

1426. Fruits and vegetables furnish how much percent of vitamin C and how much percent of vitamin A in the world:

(a) 30, 60 (b) 60 30
(c) 190, 30 **(d) 90, 60**

1427. Fruits of banana mature after how many months of fruit setting:

(a) 1-2 (b) 2-3
(c) 9-10 **(d) 3-4**

1428. Fruits with maximum iron content in:

(a) Mango (b) Date palm
(c) Apple **(d) Karonda**

1429. Function of runner is:

(a) To store reserve food (b) Propagation of plant
(c) Both these (d) None of these

1430. Function of sex cells is to:

(a) Produce flowers (b) Create next generation
(c) Both these (d) None of these

1431. Furete is an improved variety of:

(a) Avocado (b) Mango
(c) Pean nut (d) Banana

1432. 'Furete' an improved variety of avocado is a cross between:

(a) Mexican and Guatamalan type
(b) Gutamalan and West Indies type
(c) Mexican and West Indies type
(d) None.of the above

1433. GA_3 spray at which rate can increase the size of grape:

(a) 20 ppm (b) 1000 ppm
(c) 1 ppm (d) 2 ppm

1434. Gandevi Selection is a selection from which of the following?

(a) Gross Michel (b) Dwarf Cavendish

(c) Lalkela (d) Poovan

1435. Ganesh Kriti is a variety of:

(a) Ber (b) Phalsa

(c) Loquat (d) Anar

1436. 'Ganesh' a national variety of pomegranate is a selection from which of the following?

(a) Alandi/Vadki (b) Dholka

(c) Nabha (d) Kandhari

1437. Generally apple is propagated by:

(a) Seeds (b) Cuttings

(c) Grafting (d) Layering

1438. Generally litchi is propagated by:

(a) Grafting (b) Budding

(c) Cutting **(d) Layering**

1439. Genus to which grape belongs is?

(a) *Carica* **(b) *Vitis***

(c) *Punica* (d) *Citrus*

1440. Heterostyly is found in which of the following fruit crop?

(a) Litchi (b) Avocado

(c) Rambutin (d) Carambola

1441. Gibberellins enhances the sugar level of fruits through action on which of the following?

(a) Decreased amylase activity

(b) Increased amylase activity

(c) Increasing acid content

(d) None of the above

1442. Global position of mango among fruit crops is:

(a) 2 (b) **4**

(c) 3 (d) 5

1443. Global warming is attributed to increase in concentration or green house gases like:

(a) CO_2 (b) CFCs

(c) CH_4 **(d) All of these**

1444. Gola is a early variety of:

(a) Mango **(b) Ber**

(c) Papaya (d) Citrus

1445. Golden Delicious is which season variety?

(a) Mid (b) Early

(c) Late (d) Baramasi

1446. Golden Star is the variety of:

(a) Litchi **(b) Carambola**

(c) Guava (d) Rambutan

1447. Homozygosity in early generations is fast due to:

(a) Selfing (b) Back crossing

(c) Sib-mating (d) Hybridization

1448. Good average palm oil yield/ha is (tonnes/ha)

(a) 2-3 (b) 6-7

(c) 4 -5 (d) 8-9

1449. Good jelly can be prepared from:

(a) Litchi (b) Mango

(c) Guava (d) Amla

1450. Good quality of pectin is found in guava fruits which are:

(a) Green in stage **(b) Ripe but green stage**

(c) Yellow but hard stage (d) Over ripe stage

1451. Grafted mango trees become ready for transplanting:

(a) Early (b) Late

(c) Rarely (d) Never

1452. Grafting between which of the following resulted in 'Bizzare Orange' a graft chimera?

(a) Sweet orange and sour orange

(b) Sour orange and pummelo

(c) Grapefruit and sour orange

(d) Sour orange and Citron

1453. Granulation can be reduced by spraying:

(a) Lead arsenate (b) Urea

(c) Copper oxide (d) KNO_3

1454. Granulation is a disorder of which fruit crop?

(a) Mango **(b) Citrus**

(c) Sapot (d) Pomegranate

1455. Grape does not thrive in:

(a) Dry summer **(b) Humid summer**

(c) Long winter (d) Windy summer

1456. Grape fan leaf virus (GFLV) is transmitted by agency?

(a) Bacteria (b) Aphid

(c) Nematode (d) Scale

1457. Grape fruit, lemon, lime, oranges & Sweet orange are species of:

(a) *Citrus* (b) *Vitis*

(c) *Mangifera* (d) *Phyllanthus*

1458. Grape fruit are closely related to:

(a) Sweet orange **(b) Pummelo**

(c) Lime (d) Mandarin

1459. Grape guard, is a craft paper coated with which of the following:

(a) Pottassium metabisulphite and plastic polymer

(b) Sodium bisulphate and plastic polymer

(c) Sodium metabisulphate and plastic polymer

(d) None of these

1460. Grape has originated in the areas extending from Black sea and Caspian sea but was introduced in India from:

(a) Australia **(b) Afghanistan**

(c) Brazil (d) Iran

1461. Grape is a:

(a) Sub Tropical fruit (b) Temperate fruit

(c) Humid region fruit (d) All season fruit

1462. Grape rootstock resistant to Phylloxera:

(a) Freedom **(b) St. George**

(c) Harmony (d) None of these

1463. Grape tree will start fruiting after how many years of planting:

(a) 4 years (b) 5 years

(c) 1.5 years (d) 2.5 years

1464. Grape variety developed by Clonal Selection is:

(a) Pusa Seedles (b) Pusa Navrang

(c) Pusa Urvashi (d) Pusa Kanchan

1465. Grape fruit is a citrus but it is called grape fruit because

(a) Fruit taste like Grapes

(b) Fruits are borne in clusters like grapes

(c) Growing habit is like grapes

(d) Nutritional requirement is like grapes

1466. Grape fruit it commercially propagated by

(a) Seeds **(b) Cuttings**

(c) Air layering (d) Budding

1467. Grapefruit originated in:

(a) Tamil Nadu (b) MP

(c) Maharashtra **(d) Outside India**

1468. Grapes are generally dried in:

(a) Sun (b) Dehydrator

(c) Spray drier (d) Vacuum drier

1469. Grapes are pruned only once in North India while twice in the Southern India, This difference is due to the fact that:

(a) Early in bearing in Northern India white late in Southern India

(b) Bear lightly in Northern India while they bear Heavily in southern India.

(c) Late prone to disease In Northern India while they are more prone to disease in Southern India

(d) Shed leaves in winter in Northern India while they remain evergreen in Southern India

1470. Grapes contain which of the following acids?

(a) Malic acid **(b) Tartaric acid**

(c) Hydrocyanic acid (d) Glutonic acid

1471. Greenhouse gas having largest contribution to global warming;

(a) Methane (18%) (b) Nitrous oxide (5%)

(c) Carbon dioxide (60%) (d) CFC

1472. Greening In citrus is caused by

(a) Virus (b) Fungi

(c) MLOs (d) Protozoa

1473. Gross Michel is a high yielding variety of banana but susceptible to:

(a) Bunchy top **(b) Panama wilt**

(c) Black tip (d) Anthracnose

1474. Growth curve of fruit development in apple:

(a) **Single sigmoid** (b) Double sigmoid

(c) None of these (d) Triple sigmoid

1475. Growth of plants towards light to called:

(a) **Phototropism** (b) Photoperiodism

(c) Photocromatism (d) Photorespiration

1476. Growth regulator acts anti to gibberelltns:

(a) Cytokinins (b) **Paclobutrazol**

(c) Morphactin (d) Auxin

1477. Growth regulator CCC to used in grapes for

(a) **Increased vegetative growth** (b) Increases TSS

(c) Increase fruitfulness (d) None

1478. Growth regulator commercially used for control for fruit drop in citrus:

(a) NAA (b) **2,4-D**

(c) IBA (d) IAA

1479. Growth regulator that is anti gibberellins in action:

(a) **Cultar/ PP 222** (b) Cytokinin

(c) Auxins (d) Ethylene

1480. Guardling operation, winter ploughing and application of 0.45 to 0.9 kg of nitrogen per tree help mango trees

(a) **Bear regularly** (b) Remain healthy

(c) Escape disease (d) Escape insect pests

1481. Guava bears fruits mostly on:

(a) Spurs (b) **Current season Growth**

(c) Pineapple (d) Sapota

1482. Guava berry develops from:

(a) **Superior ovary** (b) Inferior ovary

(c) Intermediate ovary (d) a and b

1483. Guava contains:

(a) 300 mg/100g pulp (b) 200 mg/100g pulp

(c) 100 mg/100g pulp (d) 600mg/100g pulp

1484. Guava is a native of:

(a) Tropical south east Asia (b) Tropical Africa

(c) Tropical America (d) Sub-tropical India

1485. Guava is commercially propagated through:

(a) Whip grafting **(b) Cutting**

(c) Chip budding (d) Inarching

1486. Guava it generally trained by

(a) Open centre system (b) Central leader system

(c) Modified leader system (d) None of the above

1487. Guava is propagated

(a) Through seed (b) Vegetatively

(c) By both the above methods (d) By none of these methods

1488. Guava is the richest source oft

(a) Vitamin A (b) Vitamin B_1

(c) Vitamin B_2 **(d) Vitamin C**

1489. Guava tree will give the fruits after how many years of planting:

(a) 4 years **(b) 2 years**

(c) 7 years (d) 10 years

1490. Guava variety said to be resistant to wilt is?

(a) Allahabad Safeda (b) Sardar Guava

(c) Red Fleshed (d) Apple guava

1491. Guava wilt is caused by fungus. The variety said to be resistant to wilt is:

(a) Allahabad Safeda (b) Sardar Guava

(c) Red Fleshed (d) Apple guava

1492. Guava wilt is caused due to:

(a) *Fusarium solani* (b) *Macrophomina phaseoli*

(c) *Cephalosporium spp.* **(d) *Gliocladium reseum***

1493. Gutta Pecha is originated in

(a) South America (b) Tropical Africa

(c) China (d) None of these

1494. Gulab Jamun. *Syzygtum jambose* belongs to guava family Myrtaceae. The ploidy level of Gulab Jamun is:

(a) Tetraploid (b) Diploid

(c) Octaploid (d) Triploid

1495. Gummosis in almond is due to:

(a) Bacteria and Cu deficiency (b) Fungus

(c) Virus (d) MLOs

1496. Gummosis in citrus is caused by:

(a) Virus **(b) Fungi**

(c) MLOs (d) Nematode

1497. Gynomoecious means:

(a) Having female and bisexual flowers

(b) Having all female flowers

(c) Having male flowers only

(d) None

1498. Hachiya, Hykume and Fuyu are varieties of:

(a) Peach **(b) Persimmon**

(c) Apricot (d) Apple

1499. Halawy, Khadrawy are the varieties of:

(a) Phalsa **(b) Date**

(c) Plum (d) Jamun

1500. Hand pollination is most useful in which of the following fruits?

(a) Date (b) Grape

(c) Pineapple (d) Mango

1501. Haploid cells are produced as a result of:

(a) Meiosis **(b) Mitosis**

(c) Both these (d) None of these

1502. Hard-wood cutting is a propagating material used in:

(a) Dogwood (b) Mango

(c) Fig (d) Begonia

1503. Hawk moth is a serious pest of:

(a) Fig (b) Papaya

(c) Durian (d) Mangosteen

1504. Head system, cone system, cordon system, pergola system, single stem, avenue system and arbour system are the systems of:

(a) Planting cherry **(b) Training of grape vine**

(c) Training duranta hedge (d) Training citrus tree

1505. 'Hen and Chicken' a disorder is observed in which fruit crops?

(a) Mango (b) avocado

(c) Grape (d) Litchi

1506. Herbaceous stem cuttings propagation is used in:

(a) Carnation (b) Begonia

(c) Grape (d) Hollyhock

1507. Hesperidium is botanical form of fruit in which of the following crops?

(a) Citrus (b) Apple

(c) Mango (d) Coconut

1508. Hesperidium, Balusta and Amphisarica are the modified forms of:

(a) Pome (b) Drupe

(c) Berry (d) Nuts

1509. Hexagonal system of planting accommodates (%) more plants than square system:

(a) 15 (b) 10

(c) 20 (d) 25

1510. Hezalnut belongs to the family:

(a) Proteaceae (b) Ebenaceae

(c) Betulaceae (d) Euphorbiaceae

1511. High cytokinin/auxin ratio in callus favours:

(a) Shoot formation (b) Root development

(c) Flowering (d) Somatic embryogenesis

1512. High monsoon humidity helps in the pollination of:

(a) Custard apple (b) Aonla

(c) Phalsa (d) Guava

1513. High production in protected cultivation is due to:

(a) High photosynthetic efficiency

(b) High temperature

(c) High 0_2

(d) All of the above

1514. Highest total number of species reported in mango is 69. Who has reported this?

(a) Kosterman and Bompard (b) Mukherjee et al.

(c) Bums and Prayag (d) All of these

1515. Highly profitable intercrop in coconut and arecanut:

(a) Tapioca **(b) Pepper**

(c) Pineapple (d) Banana

1516. Hillawi is very popular variety of throughout the varieties of

(a) Grape (b) Mango

(c) Papaya **(d) None of these**

1517. Hillawi; 'Khadrawi, 'Shamran, Medgool 'Deglect Noor 'Barhee' and 'Heyani' are varieties of:

(a) Aonla (b) Custard apple

(c) Phalsa **(d) Datepalm**

1518. Himrod and Perlette are the varieties of:

(a) Grape (b) Mango

(c) Papaya (d) Guava

1519. Horse latitude lies in between North and South latitudinal belt of equator

(a) 10 to15° (b) 0 to 50°

(c) 20 to 25° **(d) 30 to 35°**

1520. Horticultural Society of India (HIS) was established in:

(a) 1942 (b) 1947

(c) 1941 (d) 1950

1521. Horticulture has been derived from" hortus" which is:

(a) German **(b) Latin**

(c) English (d) French

1522. Hot water treatment for control of anthracnose in mango is at

(a) 52 °C (b) 32°C

(c) 70°C (d) 100°C

1523. How Annona variety 'African Pride' has been developed by cross between:

(a) Sugar apple and *Annona reticulata*

(b) Sugar apple and *A. cherimoya*

(c) Sugar apple and *A. glabra*

(d) Sugar apple and *A. squamosa*

1524. How apple can be classified?

(a) Short day **(b) Long day**

(c) Day neutral (d) a & b

1525. How are isogenic lines developed through in fruits?'

(a) Mass Selection (b) **Back Cross**

(c) Mutation (d) Pure line selection

1526. How can citrus fruits be preserved for a few weeks at room temperature?

(a) Submerging them in cold water and then drying

(b) Putting them in atmosphere rich in C02

(c) By wrapping individual fruit in HDP envelops

(d) Keeping them in well ventilated room

1527. How can we easily control powdery mildew of grape?

(a) Sulfex (b) Benomyl

(c) Bavistin (d) None of the above

1528. How did M9, a dwarfing rootstock of apple has originated

(a) Hybridization (b) Mutation

(c) Chance seedling selection (d) Polyploidy

1529. How improvement in papaya is done?

(a) Inbreeding and sibmating (b) Polyploidy

(c) Hybridization (d) Mutation

1530. How is acid lime should be propagated?

(a) Seeds (b) Cuttings

(c) Air layering (d) T. Budding

1531. How areca nut is commonly propagated?

(a) By seed (b) By cuttings

(c) By Grafting (d) By Budding

1532. How is banana pesudostem called after harvesting?

(a) Desuckering **(b) Mattocking**

(c) Propping (d) Parring

1533. Ber is commercially propagated by:

(a) Seeds **(b) T-budding**

(c) Inarching (d) Layering,

1534. How is botanically Pineapple guava/Feijoa/New Zealand banana known as?

(a) *Psidium friedrichsthalianum*

(b) *Feijoa sellowiana*

(c) *P. guinenesis* (d) *P. guajava*

1535. How is crop regulation (Bahar Treatment) achieved in citrus, pomegranate and guava?

(a) Withholding of water

(b) Root and shoot pruning

(c) Deblossoming by application of NAA

(d) All the above

1536. How pineapple is usually propagated?

(a) Air layering (b) Budding

(c) Inarching **(d) Suckers**

1537. How is pollination affected in papaya?

(a) Insect (b) Wind

(c) Water (d) Animal/man

1538. How is pond apple (*Annona glabra*) used commercially?

(a) Pollinizer **(b) Rootstock for Annona**

(c) Interstem (d) Scion variety

1539. How is seedlessness is grape controlled?

(a) Dominant gene
(b) **2 complementary gene**
(c) Recessive gene
(d) Duplicate gene

1540. How is strawberry propagated through?

(a) Stolen
(b) **Runners**
(c) Crown
(d) Suckers

1541. How is the red colour of mango skin controlled?

(a) Additive gene action
(b) Dominant gene influenced by duplicate gene
(c) Cytoplasmic
(d) Trasgressive segregation

1542. How is the Cavendish banana propagated through?

(a) Seed
(b) Sucker
(c) Leaf cutting
(d) Stem cutting

1543. How is the early post blossom fruit drop in apple known?

(a) May drop
(b) June drop
(c) July drop
(d) August drop

1544. How is the inception of ripening in grape termed as?

(a) Colour break
(b) Verasion
(c) Turning stage
(d) Climacteric

1545. How is the incorporation of desirable gene from alien species into commercial cultivar without sexual mating known?

(a) Transformation
(b) Embryo rescue
(c) Tissue culture
(d) hybridization

1546. How is the Iron deficiency in plant characterized?

(a) Intervenial chlorosis of old leaves
(b) Little leaf
(c) Yellowing of new leaves
(d) Mottling of leaf

1547. How is the propagation of strawberry done?

(a) Suckers (b) **Runners**

(c) Slips (d) Stolons

1548. How long it takes mango seed lose their viability?

(a) 3-4 week (b) **4-6 week**

(c) 5-6 week (d) 6-7 week

1549. How many are the basic chromosome number (x=) mango?

(a) 20 (b) **10**

(c) 40 (d) 80

1550. How many plants can be accommodated through high density planting in pineapple?

(a) 10,000 (b) 45,000

(c) 30,000 (d) **60,000**

1551. How many plants of Amrapali Mango can be planted in one hectare?

(a) 100 (b) 400

(c) 1600 (d) 1000

1552. How micro propagation provides disease free plants?

(a) Embryo rescue (b) **Meristem culture**

(c) Micro propagation (d) Anther culture

1553. How much is the average yield (kg/ha) of arecanut?

(a) 2000-2500 (b) 500-700

(c) >3000 (d) **1250-1600**

1554. How much is the flower set from fruit to maturity in guava?

(a) 0.1-0.5% (b) **35-55%**

(c) 5-15% (d) 80-85%

1555. How much is the good average palm oil yield/ha is (tonnes/ha).

(a) 2-3 (b) 6-7

(c) 4-5 (d) 8-9

1556. How much is the per cent progeny which is seedless as a result of seedless x seedless cross followed by embryo rescue?

(a) 10 (b) 20

(c) 80 (d) 40

1557. How much is the vitamin a content in mango?

(a) 200 IU/100g (b) 1500 IU/100g

(c) 4000IU/100g (d) 3000 IU/100g

1558. How Sweet orange is commercially propagated?

(a) Seed **(b) T-budding**

(c) Hardwood cutting (d) Inarching

1559. How the guava variety Arka Amulya & Arka Mridula have been developed through?

(a) Selection (b) Mutation

(c) Hybridization (d) Polyploidy

1560. How the variety Loose Perlette a strain of Perlette cultivar of grape has been evolved?

(a) Selection (b) Mutation

(c) Polyploidy (d) Hybridization

1561. How was Red Sitaphal a variety of Annona evolved?

(a) Mutation (b) Introduction

(c) Hybridization **(d) Chance seedling**

1562. Howk Moth is a serious pest of which of the following fruit crops?

(a) Mangosteen (b) Fig

(c) Durian (d) Papaya

1563. Hazelnut/filbert/cobnut (*Corylus avellana*) belongs to family

(a) Ebenaceae (b) Proteaceae

(c) Betulaceae (d) Juglandaceae

1564. Hybrid variety of avocado is:

(a) Pollock **(b) Fuerte**

(c) Lula (d) Yulia

1565. Hybrid variety of chiku is:

(a) CO-1 (b) CO-2

(c) Bangalore (d) Badam

1566. IAA was discovered by a dutch scientist in 1934:

(a) Mullar **(b) First Went**

(c) Nelson (d) Kurosawa.

1567. Ideal pH for orchard crops is:

(a) 6-8 (b) 5-7

(c) 4-6 (d) 9-13

1568. Hybridization between which of the following resulted in 'Sampson' and Thornton' varieties of the grapefruit.

(a) Lime and Kumquat **(b) Duncan and Dancy**

(c) Stasuma and Ruby (d) Duncan and Satsuma

1569. IC_2 and 519, triploid banana hybrids have originated from the cross between:

(a) Gros Michel and Pisang Lilin

(b) Gross Michel and *M. acuminata*

(c) Gros Michel and Kadali

(d) None of the above

1570. Identify the growth phase during which seed maturation takes place?

(a) 2nd rapid phase **(b) lag phase**

(c) First phase (d) After Ripening

1571. Immunity to panama disease in banana is shown by:

(a) Cercospora (b) Fusarium

(c) Basrai dwarf (d) Synchytrium

1572. Important disease of banana in India is:
(a) Panama wilt (b) **Bunchy top**
(c) Cercospora leaf spot (d) Sigatoka

1573. Improved Golden Yellow and California Advance are varieties of which fruit?
(a) Mandarin (b) **Loquat**
(c) Apple (d) Papaya

1574. Improvement in papaya is done mainly through:
(a) Inbreeding and sib-mating (b) Polyploidy
(c) Hybridization (d) Mutation

1575. In a RBD experiment having 9 treatments and 4 replications, the error degree of freedom will be:
(a) 24 (b) 36
(c) 32 (d) 27

1576. In apple dwarfing, rootstock is good for.
(a) Low chilling (b) Scab resistance
(c) Fruit colour **(d) High density planting**

1577. In apples small, narrow and mottled leaves are primarily because of the deficiency of.
(a) Zinc (b) Iron
(c) Potassium (d) Iodine

1578. In ber best time of pruning in Northern India is:
(a) Mid March to Mid April **(b) Mid May to Mid June**
(c) Mid July to Mid August (d) Mid Sept to Mid Oct

1579. In Bombay-Deccan region, orange trees are given "root exposure treatment" to make them:
(a) Drought resistant
(b) Frost resistant
(c) Pest free
(d) Properly flower at appropriate time

1580. In budding and grafting the union between stock and scion takes place through which of the following?

(a) Phloem (b) Cortex

(c) Cambium (d) Xylem

1581. In cutting how much inches long part of one year age should be used:

(a) 6-9 (b) 12-16

(c) 1-2 (d) 15-20.

1582. In double helical structure of DNA, distance of one turn is:

(a) 36 A° (b) 38A°

(c) 34 A° (d) 3.4 A°

1583. In early pre-bearing life of pear orchards, peach is grown in between pear plants because it:

(a) Generate some additional Income

(b) Protect pear trees from hot wind

(c) Improves microclimate for pears

(d) Prevents sunscald of pear fruit

1584. In general, fruits and vegetables are not a rich source of.

(a) Vitamin C (b) Vitamin B,

(c) Vitamin A (d) Minerals

1585. In grafting lower part of plant is know as:

(a) Bud (b) Apical

(c) Stock (d) Scion

1586. In grafting upper part of stem is called:

(a) Stock **(b) Scion**

(c) Bud (d) Apical

1587. In guava high temperature at the time of fruit development causes

(a) Guava wilt **(b) Fruit drop**

(c) Yellow leaf disease (d) Guava canker

1588. In guava which method is more prevalent and successful in comparison to other methods of vegetative propagation:

(a) **Inarching** (b) Forkert budding

(c) Shield budding (d) Root cutting

1589. In heterosis breeding, one should capitalize on:

(a) In breeding depression (b) **Over dominance**

(c) Higher seed vigour (d) Homozygosity

1590. In hexagonal method of tree planting, about how many percent more trees can be planted than triangular method,

(a) 25 (b) 40

(c) 10 (d) **15**

1591. In hilly area which system of fruit planting should be used:

(a) **Contour** (b) Strip cropping

(c) Wind (d) Strib

1592. In India Acid lime is generally used as:

(a) Jam (b) Pickle

(c) Juice (d) **Pickle & Juice**

1593. In India leading lime production state it:

(a) Bihar (b) Karnataka

(c) Assam (d) **AP**

1594. In India maximum area under litchi is in:

(a) Gujarat (b) Karnataka

(c) Maharashtra (d) **Bihar**

1595. In India, banana contributes following percentage of fruit production:

(a) **80%** (b) 60%

(c) 31% (d) 10%

1596. In India, banana was introduced in Bengal by the end of:

(a) 11th Century (b) 12th Century

(c) 17th Century (d) 19th Century

1597. In India, the leading state in area of mango it:

(a) Uttar Pradesh (b) Bihar

(c) AP (d) WB

1598. In India, a chiku tree yields fruits:

(a) Once a year **(b) Twice a year**

(c) Thrice a year (d) Fourth-times a year

1599. In Navel Orange, navel is present at the:

(a) Microphyler end (b) Styler end

(c) Thalamic end (d) None of these

1600. In Nilgires the main crop of mandarin Orange is harvested in:

(a) July **(b) August-October**

(c) December-September (d) March

1601. In North India cuttings of grapes are taken from one-year- old wood vines when they are

(a) Flowering (b) Actively growing

(c) Dormant (d) Bearing fruits

1602. In North India grape gives one crop in a year during:

(a) Winter (b) Rainy season

(c) Summer (d) Autumn

1603. In North India the ripening of grapefruit take place during

(a) Oct-Nov **(b) Nov-Jan**

(c) Dec-Jan (d) Jan-Fed

1604. It North India, rooted cuttings of grapes are transplanted in January and February when they are:

(a) Dormant (b) Actively growing

(c) Just producing new leaves (d) Beaming flower-buds

1605. In North-India, the earliest ripening peach variety is:

(a) Flordasun
(b) Pratap
(c) Shan-e-Punjab
(d) Sharbati

1606. In orchard layout smallest area should be allocated under

(a) Irrigation channels
(b) Roads and path
(c) Manures pit
(d) Farm building

1607. In papaya, pollination is affected by:

(a) Insect
(b) Wind
(c) Water
(d) Animal/ Man

1608. In peaches the traditional system of training is.

(a) open center
(b) Multiple stem
(c) Central leader
(d) Modified central leader

1609. In peaches, the most serious pest is:

(a) Leaf curl aphid
(b) Black aphid
(c) Case worm
(d) Defoliating beetle

1610. In Punjab only successful cultivated plum is:

(a) American
(b) Chinese
(c) Japanese
(d) Indian

1611. In situ method of mango grafting

(a) Epicotyl grafting
(b) Inarching
(c) Soft wood grafting
(d) Stooling

1612. In Southern India grape vines bear fruits in:

(a) October-November
(b) January-February
(c) December
(d) April and August-September

1613. **In the presence of sunlight, CO_2 and H_2O (with the help of chlorophyll) are converted into carbohydrates, this is called as:**

(a) Respiration (b) Solar radiation

(c) Metabolism (d) **Photosynthesis**

1614. **In which concentration sodium benzoate is added for the preservation of fruit juice?**

(a) 0.0/1-0.5% (b) **0.06-0.10%**

(c) 1.1-1.5% (d) 1.6-2.0%

1615. **In which crop dichogamy is considered to be the main cause of poor yield?**

(a) Loquat (b) **Custard apple**

(c) Strawberry (d) Plum

1616. **In which fruit the bract, perianth and seed are edible**

(a) Pear (b) **Jack fruit**

(c) Banana (d) Walnut

1617. **In which groups can the seed-less variety of guava placed?**

(a) Haploid (b) **Anueploid**

(c) Diploid (d) Tetraploid

1618. **In which of the cell divisions the chromosome number in the nuclei of the resulting daughter cells is reduced half?**

(a) **Meiosis** (b) Mitosis

(c) Both these (d) None of these

1619. **In which of the following cultivar of mango Neelum is not a parent?**

(a) Ratna (b) Mallika

(c) **Basti-5** (d) Amrapali

1620. **In which of the following fruit crops pollen sterility is major problem?**

(a) **Ber** (b) Grape

(c) Aonla (d) Pomegranate

1621. In which of the following fruit follows a double sigmoid curve?

(a) Mango **(b) Fig**

(c) Apple (d) Pear

1622. In which of the following fruits does vegetative parthenocarpy takes place?

(a) Peach (b) Bael

(c) Ber **(c) Banana**

1623. In which of the following fruits is self- barrenness practically universal?

(a) Cherry **(b) Apricot**

(c) Apple (d) Almond

1624. In which of the following fruits losses during transit are more

(a) Banana (b) Ber

(c) Guava (d) Apple

1625. In which of the following fruits the mesocarp is edible?

(a) Apple (b) Cashewnut

(c) Papaya (d) Jackfruit

1626. In which of the following spongy tissue developed in the mesocarp of the fruit as a physiological disorder?

(a) Guava **(b) Mango**

(c) Banana (d) litchi

1627. In which of the following operation tip, simple, trench, mound, pot and air are used as prefixes?

(a) Tillering **(b) Layering**

(c) Notching (d) Grafting

1628. In which of the fruit growing region Chestnut is grown?

(a) Temperate (b) Tropical

(c) Semi and (d) Sub tropical

1629. In which of the organic acid tamarind (*Tamarindus indica*) is rich in?

(a) Tartaric (b) Malic
(c) Citric (d) None of these

1630. In which part of guava fruit Ascorbic acid content is maximum?

(a) Pulp **(b) Peel**
(c) Seed (d) Core

1631. In which region Bahar treatment in citrus is practiced?

(a) East & North India **(b) Central & West India**
(c) North & East India (d) West & South India

1632. In which region did citrus originated?

(a) South Africa **(b) South-East China**
(c) West Indies (d) America

1633. In which year Central Tuber Crop Research Institute was established?

(a) 1963 (b) 1964
(c) 1987 (d) 1806

1634. In which year National Bureau of Plant Genetic Resources was established.

(a) 1978 (b) 1986
(c) 1976 (d) 1985

1635. In world 90% of grapes produced are used as:

(a) Table (b) Raisin
(c) Juice **(d) Wine**

1636. In world, from the Litchi production point of view, India ranks:

(a) First (b) Second
(c) Third (d) Fourth

1637. In young pomegranate fruits cracking is due to:

(a) Zinc deficiency (c) Sulphur deficiency

(c) Boron deficiency (d) Molybdenum deficiency

1638. Inarching in mango in India was suggested by:

(a) H Lynch (b) R. N Singh *et. al*

(c) S. K. Mukherjee (d) R.M. Pandey *et. al.*

1639. Inarching is practiced to repair damaged:

(a) Root (b) Shoot

(c) Both (d) None of these

1640. Increase in ethylene production with ripening is characteristics of:

(a) Citrus fruits (b) Grapes

(c) Climacteric fruits (d) Non-climacteric fruits

1641. India contribute following percentage of mango production in the world:

(a) 20% (b) 10%

(c) 56% (d) 80%

1642. India is third largest producer of coconut next only to :

(a) China and Brazil (b) Srilanka and Maldives

(c) Philippines and Indonesia (d) Japan and Korea

1643. Indian council of Agricultural Research (ICAR) is an;

(a) Autonomous registered Society

(b) Agricultural Research Organization of Govt. of India.

(c) National Department of Agriculture

(d) DST. Govt. of India

1644. Indian Institute for Horticultural Research (IIHR), Bangalore was established:

(a) 1967 (b) 1969

(c) 1968 (d) 1990

1645. Indian Journal of Horticulture is a publication of:

(a) ICAR (b) IARI

(c) Horticultural Society of India (d) CAIR

1646. Indicator plant for tristeza virus in citrus:

(a) *Citrus reticulata* **(b) *C. aurantifolium***

(c) *C. paradisica* (d) *C. limmetoides*

1647. Indole-3-acetic acid is a/an

(a) Plant hormone (b) Insecticide

(c) Enzyme (d) Organic acid

1648. Indole butyric acid is a:

(a) Root promoting substance

(b) Shoot promoting substance

(c) Flowering inducing substance

(d) Growth retarding substances

1649. Inferior types of bananas also grow in:

(a) Desert (b) Sea

(c) Himalayas (d) Rivers

1650. Inarching is practiced to repair the roots whereas to repair stem which type of grafting is used?

(a) Bridge grafting (b) Approach grafting

(c) Bench grafting (d) Book grafting

1651. Insect responsible for carprification is:

(a) Inderbala spp. **(b) Blastophagapsens**

(c) Fruit fly (d) Butterfly

1652. International Institute of Horticulture is situated in:

(a) Brazil (b) Italy

(c) USA (d) Norway

1653. International Society for Horticultural Science (ISHS) has HQ in:

(a) The Netherlands
(b) Belgium
(c) France
(d) New York

1654. Interveinal chlorosis is caused by deficiency of

(a) Nitrogen
(b) Magnesium
(c) Potassium
(d) Calcium

1655. Introduction of Kinnow in India was done in:

(a) 1940
(b) 1959
(c) 1970
(d) 1985

1656. Invaders from Afghanistan and Persia introduced grapes in to India in:

(a) 1600 AD.
(b) 1300 AD
(c) 1400 AD
(d) 1500 AD

1657. Iron deficiency in plant is characterized by:

(a) Interveinal chlorosis of old leaves
(b) Little leaf
(c) Yellowing of new leaves
(d) Mottling of leaf

1658. It has been observed that the malta sweet orange budded on rough lemon produced trees how many times larger than those budded on citrus (*C. medica*),

(a) Two
(b) Three
(c) Four
(d) Five

1659. Jack fruit is commercially propagated by:

(a) Seed
(b) Inarching
(c) Budding
(d) Layering

1660. Jackfruit (*Artocarpous heterophyllus*) breadfruit (*A. alfilis*), Monkey jackfruit (*A. lakoocha*) belong to family Moraceae. The Jackfruit is a tetraploid its somatic chromosome number (2n) is:

(a) 50
(b) 56
(c) 34
(d) 40

1661. Jackfruit belong to family:

(a) Carcaceae (b) Anacardiaceae
(c) Myrtaceae **(d) Moraceae**

1662. Jaffa sweet orange is the best variety grown in:

(a) Haryana **(b) Punjab**
(c) HP (d) AP

1663. Jalore seedless a pomegranate variety is recommended for cultivation in the state of:

(a) Rajasthan (b) MH
(c) Gujarat (d) HP

1664. Jam comes under:

(a) Concentrate (b) **Sugar concentrate**
(c) Any other (d) Non heat processing

1665. Jameri, Jatti, Khatti and Kharna Khatta provide suitable root-stocks for:

(a) Sweet orange (b) Sweet lime
(c) Grapes (d) Pomegranate

1666. Jam is perepared with:

(a) Fruit pectin and Pulp
(b) Fruit pulp alone
(c) Fruit pectin alone
(d) Pectin, pulp, seed and skin mix

1667. Jamun also called as Java Plum (*Syzigium Cuminii*) belong to family Myrtaceae. The seeds of java plum are:

(a) Polyembryonic (b) Monoembryonic
(c) Seedless (d) Pathenogenetic

1668. Jamun belongs to same family as:

(a) Guava (b) Peach
(c) Apricot (d) Jackfruit

1669. Jargonelle, an improved variety of
   (a) Almond (b) Plum
   (c) Pear (d) Apple

1670. Triplod cultivars of apple
   (a) Baldwin (b) Gravenstin
   (c) Winesap **(d) All**

1671. JH Hale peaches are:
   (a) Male sterile (b) Self fruitful
   (c) Male fertile (d) Parthenocarpic

1672. Jonathan is variety of:
   (a) Apple (b) Guava
   (c) Grape (d) Avocado

1673. June drop" in apple is mainly due to:
   (a) Lack of pollination (b) Alternate bearing
   (c) Moisture Stress (d) Calcium deficiency

1674. Colour mutant cultivar of apple
   (a) Baldwin (b) Gravenstin
   (c) Michal **(d) Skyline supreme**

1675. Apricot which variety well suitable as a kitchen garden
   (a) Nugget **(b) Moorpark**
   (c) Kaisa (d) Shakarpara

1676. Juvenile phase of plant growth is predominantly:
   (a) Reproductive phase **(b) Vegetative growth phase**
   (c) Both these (d) None of these

1677. Which is cultivar of peach
   (a) Baldwin **(b) Peshwari**
   (c) Winesap (d) All

1678. Kiwi fruit firstly planted in

(a) **Lal Bagh, Bangalore** (b) Botanical garden, Kolcutta

(c) FRS, Pune (d) None

1679. Passion fruit is a

(a) **Perennial climber** (b) Annual plant

(c) Perennial plant (d) None

1680. Juicy pulpy "Aril" is edible part of

(a) Mango (b) Apple

(c) **Rambutan** (d) Banana

1681. 'Jyoti' pomegranate is colonel selection from

(a) Kandhari (b) Alandi

(c) **Ganesh** (d) Beesein Seedless

1682. Kagzi lime is the best indicator for.

(a) Leaf curl (b) **Tristeza**

(c) Xyloporosis (d) Exocortis

1683. Kahir Safeda Guava is a cross between;

(a) Allahabad Safeda and Kohir

(b) L -49 and Allahabad Safeda

(c) **Kohir and Allahabad Safeda**

(d) L-49 and Lalit

1684. Kalapady is a dwarfing and polyembryonic rootstock for:

(a) **Mango** (b) Jamun

(c) Citrus (d) None of these

1685. Kanchan and Krishna are varieties of :

(a) Grape (b) Mango

(c) **Aonla** (d) Plum

1686. Kagzi Nimbu (Sour lime) is propagated mainly by:

(a) **Seed** (b) Cuttings

(c) Grafting (d) Budding

1687. Karaka, Nagpur, Katha, Mehrun, Aligani and Jogia are the varieties of:

(a) **Ber** (b) Aonla

(c) Wood Apple (d) Apple

1688. Karonda has originated from:

(a) **India** (b) South Africa

(c) Srilanka (d) America

1689. Kaveri is a hybrid variety of:

(a) **Passion fruit** (b) Amla

(c) Butter fruit (d) Loquat

1690. Kaveri was developed at IIHR, Bangalore. The parents of Kaveri are:

(a) **Purple & yellow passion fruit**

(b) Yellow & green passion fruit

(c) Red & yellow passion fruit

(d) None of these

1691. Keiffer and Le Conte pears are the result of cross between.

(a) ***Pyrus cummunis & P. pyrifolia***

(b) *P. serolina* & *P. cummunis*

(c) *P. pyrifolia* & *P. pashia*

(d) None of these

1692. Kera shree and Laksh Ganga are varieties of

(a) **Coconut** (b) Guava

(c) Apple (d) Cashewnut

1693. Kew and Giant Kew varieties of pineapple belong to:

(a) Spanish Group (b) **Cayenne group**

(c) Queen group (d) Abacaxi group

1694. Kew Queen and Mauritius' are the popular varieties of

(a) Apple (b) Cherry

(c) Pear (d) **Pineapple**

1695. Kharna Khatta. Jamberi and Rabab tenga are:

(a) Varieties of grapefruit

(b) Varieties of lime

(c) Used at rootstocks for grafting of grape fruits

(d) Varieties of sweet oranges

1696. Khasi mandarin is the best variety in:

(a) Gujarat (b) **Assam**

(c) Punjab (d) Karnataka

1697. Kinnow cultivar of citrus is a hybrid between:

(a) King x Willow leaf (b) King x Owari

(c) King x Cleopatra (d) KingxDancy

1698. Kinnow is a cross between:

(a) *Citrus nobilis* and *C. deliciosa*

(b) *C. nobilis* and *C. reticulata*

(c) *C. reticulata* and *C. sinenses*

(d) *C. nobilis* and *C. sinensis*

1699. Kinnow is a famous variety of mandarin in:

(a) Karnataka (b) Maharashtra

(c) Tamil Nadu (d) **Punjab**

1700. Kinnow is a basically a:

(a) Tangerine (b) **Tangor**

(c) Mandarin (d) Citrange

1701. Kinnow mandarin was developed by

(a) **HB Frost** (b) Swingle
(c) Spigeal Ray (d) Tanaka

1702. Kinnow mandarin was introduced in India in

(a) 1947 (b) 1950
(c) **1959** (d) 1965

1703. Kniffin system of training in grape was developed by Mr. William Kniffin in:

(a) Rome (b) **New York (USA)**
(c) India (d) France

1704. Kniffin system of training is used in:

(a) Litchi (b) Papaya
(c) Mango (d) **Grapes**

1705. Kohir Safed and Safed jam are the varieties of

(a) Pomegranate (b) **Guava**
(c) Date palm (d) Mango

1706. Krishna variety of aonla is good for:

(a) Jam (b) Jelly
(c) **Murabba** (d) Pickles

1707. L-49 is a popular variety of;

(a) Pear (b) Peach
(c) Litchi (d) **Guava**

1708. L-49 is a selection from Allahabad Safeda made during 1927 by Dr. GS Cheema at:

(a) **Pune** (b) Allahabad
(c) Lucknow (d) New Delhi

1709. Laksha Ganga' is an improved hybrid of which of the following?

(a) Aracanut (b) **Coconut**
(c) Pecan not (d) Walnut

1710. Lal Ambri is the first generation hybrid of which one of the following

(a) Ambri x Red delicious **(b) Golden delicious x Ambri**

(c) Ambri x Golden delicious (d) Red delicious x Ambri

1711. Lalit is an improved variety of:

(a) Guava (b) Papaya

(c) Grape (d) Pineapple

1712. Langra mango variety is through:

(a) Mass selection (b) Bulk selection

(c) Chance seedling selection (d) Hybridization

1713. Largest guava producing state in India is;

(a) Punjab **(b) UP**

(c) MH (d) KN

1714. Largest area under banana in India is in:

(a) TN (b) Punjab

(c) Assam (d) WB

1715. Largest area under citrus in India is in:

(a) MP, Tamil Nadu, Maharashtra

(b) Rajasthan, Gujarat

(c) UP, Punjab, Haryana

(d) Assam, Sikkim. Meghalaya

1716. Largest fruit producing state of India;

(a) MH (b) AP

(c) Gujarat (d) UP

1717. Largest producer of citrus in the world is;

(a) India (b) USA

(c) China (d) Brazil

1718. Largest producer of fig in India (State):

(a) **MH** (b) KN

(c) UP (d) MP

1719. Largest producer of lime in the world is:

(a) India (b) China

(c) Brazil (d) South Africa

1720. Largest producer of pear in the world is:

(a) Spain (b) India

(c) Italy (d) Brazil

1721. Lasoda, *Cordia mixa* is a acid less and minor fruit of India. It belongs to family:

(a) Anacardiaceae (b) **Boraginaceae**

(c) Rosaceae (d) Euphorbiaceae

1722. Late maturing variety of ber is:

(a) Umran (b) Gola

(c) Chonchal (d) Rashmi

1723. Late variety of loquat is:

(a) Golden yellow (b) Safeda

(c) Fire ball **(d) Tanaka**

1724. Lateral shoots of branches produced at the base of main stem are called:

(a) Stolons (b) Suckers

(c) Offsets or offshoots (d) Runners

1725. Latest hybrid mango variety, released from Maharashtra is:

(a) Amrapali (b) Mallika

(c) Ratna (d) Nileshwar Dwarf

1726. Latex from which of the following is used to obtain Chicklet?

(a) Sapota (b) Fig

(c) Papaya (d) Mango

1727. Law of Minimum was given by:

(a) Shelford **(b) Liebig's**

(c) Clement (d) Blackman

1728. Layering is method of:

(a) Fruit preservation **(b) Plant propagation**

(c) Both these (d) None of these

1729. Leading country in production of apple is:

(a) South America (b) Brazil

(c) India **(d) China**

1730. Leading dollar-earning nut crop is:

(a) Mango (b) Banana

(c) Cashewnut (d) Apple

1731. Leading State in 'ber' production in India is:

(a) Rajasthan (b) Bihar

(c) Maharashtra **(d) MP**

1732. Leading state in tea production:

(a) Karnataka **(b) Assam**

(c) Tamil Nadu (d) Kerala

1733. Leaf curl in peach is caused by:

(a) Virus and bacteria (b) Insect and bacteria

(c) Micronutrient its deficiency **(d) Insect and fungus**

1734. Leaf scorching in mango is due to:

(a) K-deficiency (b) N-deficiency

(c) Chloride deficiency **(d) Chloride toxicity**

1735. Leaf spot disease of Chiku is caused by

(a) Bacteria (b) Virus

(c) Fungus (d) Mycoplasma

1736. Leaves varieties are generally:

(a) Tree (b) Semi-tree

(c) Bush **(d) Shrub**

1737. Lemon and rhododendron are propagated by:

(a) Leaf cutting (b) Leaf bud cutting

(c) Root cutting (d) All these

1738. Lemon is commercially propagated by:

(a) Seed (b) Cutting

(c) Budding (d) None of these

1739. Lemon squash is preserved for a longer period without the discoloration by use of:

(a) Malic acid and potassium metabisulphite

(b) Citric acid and sodium

(c) Benzoic acid and citric and

(d) Potassium metabisulphite

1740. Lemons are good source of:

(a) Vitamin A (b) Nicotenic acid

(c) Vitamin C (d) Iron

1741. Lime is a good source of:

(a) Vitamin A **(b) Vitamin C**

(c) Vitamin (d) Vitamin B2

1742. Linkage between genes affects:

(a) Vernalization (b) Fertilization

(c) Anaphase **(d) Independent assortment**

1743. Litchi and pistachio nut flowers are devoid of:

(a) Corolla (b) Stamen

(c) Calyx (d) Pistil

1744. Litchi belongs to the family:

(a) Rosaceae (b) Rutaceae

(c) Moraceae **(d) Sapindaceae**

1745. Litchi does not flower is south Indian plateau in winter months:

(a) For want of sufficient low temperature in winter months

(b) Because of the high temperature prevalent at the time of flower formation

(c) Because of high relative humidity present at the time flower formation

(d) Because of the excessive light stimulus present at the time of flower formation

1746. Litchi fruit is botanically:

(a) Single-seeded nut (b) Pome

(c) Single-seeded berry (d) Stone

1747. Litchi fruit start cracking, a serious disorder when summer temperature is beyond

(a) 30°C (b) 35°C

(c) 40°C (d) 45°C

1748. Litchi is commercially propagated by:

(a) Tip layering **(b) Air layering**

(c) March grafting (d) Veneer grafting

1749. Litchi is native to:

(a) North China (b) North America

(c) South China (d) West Indies

1750. Litchi leaf scorching is associated with deficiency of:

(a) N (b) P

(c) Ca **(d) K**

1751. Litchi variety less susceptible to sun burn and fruit cracking is

(a) Calcutta (b) Bombay

(c) Muzaffarpur (d) Rose scented

1752. Litchi was introduced to India from:

(a) Indonesia **(b) Japan**

(c) China (d) Malaya

1753. Little leaf in mango is associated with the deficiency of:

(a) Fe **(b) Zn**

(c) Cu (d) Mn

1754. Loquat belongs to the family:

(a) Sapotaceae **(b) Rosaceae**

(c) Sapindaceae (d) None of these

1755. Loquat fruit is a:

(a) Pome (b) Stone

(c) Berry (d) Nut

1756. Loquat is commonly propagated by:

(a) Seeds (b) Suckers

(c) Grafting (d) Stooling

1757. Loquat is native to:

(a) China (b) Brazil

(c) India (d) Tropical America

1758. Loquat belongs to the family:

(a) Rosaceae (b) Rutaceae

(c) Moraceae (d) Myrtaceae

1759. Loquat is popular fruit of:

(a) North India (b) South India

(c) East India (d) West India

1760. Loquat is strictly

(a) Tropical crop **(b) Sub-tropical crop**

(c) Temperate crop (d) None of the above

1761. Loquat variety Thomes Pride' is an introduction from:

(a) **Japan** (b) Australia

(c) China (d) South Africa

1762. Low chilling varieties of apple:

(a) Bartlett, Max.

(b) Vered, Michel, Schlomit, Winter Banana

(c) Prima, Priscella, Liberty, Freedom, Red Free

(d) Red Delicious, Golden Delicious, Royal Delicious

1763. Low-chill peach variety is:

(a) Kanto-5 (b) Shimizo Hakuto

(c) Shan-e-Punjab (d) July Elberta

1764. Low-Chill plum variety is:

(a) Santa Rosa **(b) Titron**

(c) Beauty (d) Mathley

1765. Isogenic lines are developed through:

(a) Mass Selection (b) **Back Cross**

(c) Mutation (d) Pure line selection

1766. Luckhnow-40" Allahabad Safeda' and 'Seedless' are verities of guava having:

(a) Pink flesh **(b) White flesh**

(c) Painted flesh (d) Bright red skin

1767. Lucknow - 49" variety of guava is tolerant to:

(a) Guava wilt (b) Fruit drop

(c) Stem canker (d) Anthracnose

1768. Lye peeling is done by:

(a) 10% hot acid solution (b) 10% cool solution

(c) 1% cool alkali solution **(d) 1% hot alkali solution**

1769. M9, a dwarfing rootstock of apple has originated as a result of:

(a) Hybridization (b) Mutation

(c) Chance seedling selection (d) Polyploidy

1770. Maayan, Tropical Beauty, Winter Banana, Vered, Michael, Schlomit varieties of apple belongs to which of the following groups?

(a) Low chilling apple (b) High chilling

(c) Apple scab resistant (d) None of the above

1771. Macadamia nut belongs to the family:

(a) Proteceae (b) Rosaceae

(c) Euphorbiaceae (d) Rutaceae

1772. Macadomia nut (*Macadamia termifolia*), belongs to family proteaceae. It was introduced in India from:

(a) Australia **(b) USA**

(c) Japan (d) China

1773. Madhubindu' 'Washington' 'Ceylon' 'Singapore' 'Coorg Honey Dew' and CO-1 are varieties of

(a) Papaya (b) Banana

(c) Grape (d) Orange

1774. Mahal used as rootstock of pear is botanically called as:

(a) *Pyrus pashia* (b) *P. pyrifolia*

(c) *P. serotina* (d) *P. communis*

1775. Main crop of sour lime comes from first flush which is harvested in:

(a) August **(b) April**

(c) June (d) Sep-Oct

1776. Main method of propagating sweet lime is:

(a) Seed (b) Budding

(c) Hardwood cuttings (d) Root cuttings

1777. Major bottleneck in breeding of banana:

(a) Pollen sterility (b) Non-receptivity of stigma

(c) Vegetative parthenocarpy (d) All of the above

1778. Major nut producing state in India is:

(a) Punjab (b) HP

(c) UP (d) J&K

1779. Major physiological disorder in litchi is:

(a) Black tip **(b) Fruit cracking**

(c) Soft nose (d) Storage rot

1780. Major processed product of ripe guava is:

(a) Jam (b) Nectar

(c) Jelly (d) Squash

1781. Majority of coconut roots are concentrated within a diameter of:

(a) 6 m (b) 2 m

(c) 8 m **(d) 4 m**

1782. Male and female gametes of plants originate in:

(a) Flowers **(b) Pith**

(c) Apical meristem (d) All these

1783. Male sterile peach variety:

(a) July Elberta (b) Kanto-5

(c) JH Hale (d) Shan-e-Punjab

1784. Maleness in fruits can be induced by the application of:

(a) Cycocel **(b) GA_3**

(c) IBA (d) Ethephon

1785. Mallika is a cross between:

(a) Dashehari x Langra (b) Dashehari x Neelam

(c) Dashehari x Alphonso **(d) Neelam x Dashehari**

1786. Mallika, evolved at the IARI, New Delhi, is a variety of:
(a) Guava (**b) Mango**
(c) Papaya (d) Lime

1787. Malus robusta and *M. Zumi* are source of resistance against:
(a) Powdery mildew (b) Scab
(c) Blight (d) Canker

1788. Mandarin are commercially propagated by:
(a) Layering (b) **T-budding**
(c) Tongue grafting (d) Cutting

1789. Mangiferin is:
(a) Anti-Malformin (b) Malformin
(c) Flowering agent (d) None of these

1790. Mangla' is an improved variety of which of the following crops?
(a) Coffee **(b) Arecanut**
(c) Tea (d) Coconut

1791. Mango and banana are:
(a) Non-climacteric fruits
(b) Climacteric fruits
(c) Nature of ripening depends upon varieties
(d) None

1792. Mango and citrus produce seedlings:
(a) Parthenocarpic (b) Parthenogenesis
(c) Polyembryonic (d) All of these

1793. Mango beverages are prepared by using:
(a) Colour and essence (b) Colour alone
(c) Essence alone (d) **None of the above**

1794. Mango fruit can be best stored at:

(a) 4°C **(b) 8°C**
(c) 16°C (d) None of above

1795. Mango fruits are good source of:

(a) Vit. A (b) Vit. C
(c) Riboflavin (d) Pectin

1796. Mango inflorescence contains which type of flowers:

(a) Male and hermaphrodite
(b) Male and female hermaphrodite
(c) Male and neutral
(d) Female and hermaphrodite

1797. Mango is allopolyploid and probably amphidiploid. This hypothesis was given by Dr. SK Mukherjee. Which of the following evidences support his hypothesis? (1) High number of nuclear chromosome (2) Secondary association of bivalents (3) Regular pairing and absence of multivalent formation (4) Good pollen fertility.

(a) 1 and 2 only (b) 1, 2 and 3 only
(c) 2 and 3 only **(d) All of the above**

1798. Mango is native of:

(a) South Africa **(b) Indo-Burma**
(c) China (d) Brazil

1799. Mango is stored at temperature (°C):

(a) 6-7 (b) 7-8
(c) 8-9 (d) 9-10

1800. Mango malformation is due to:

(a) Fungus (b) Bacterium
(c) Boron deficiency **(d) Air pollution**

1801. Mango malformation is reduced by application:

(a) Abscisic acid (b) GA3 @ 10 ppm

(c) NAA @ 200 ppm (d) 2,4-D @ 20 ppm

1802. Mango seed loose their viability within:

(a) 3-4 week **(b) 4-5 week**

(c) 5-6 week (d) 6-7 week

1803. Mango tree bears fruits on:

(a) Last year growth (b) Current season growth

(c) Both a & b (d) None

1804. Mango variety least affected with powdery mildew

(a) Lal Sundhari (b) Amrapali

(c) Alphonso (d) Bombay Green

1805. Mango variety most suitable for high density planting:

(a) Amrapali (b) Chausa

(c) Dashaheri (d) Langra.

1806. Mango variety resistant to mango malformation:

(a) Bhadauran (b) Bombay Green

(c) Dashehari (d) Langra

1807. Mango, papaya and dates fruit contain fair amount of which vitamin/nutrient?

(a) Vit A (b) Vit D

(c) Cit C (d) Iron

1808. Mangosteen (*Garcinia mangostana*) belongs to the family:

(a) Sapoteaceae (b) Proteaceae

(c) Guttiferae (d) Betulaceae

1809. Sapota (*Acharus sapota*) belong to the family:

(a) Sapindaceae (b) Guttiferae

(c) Sapotaceae (d) Rosaceae

1810. Mangosteen has originated from:

(a) Indo-Burma (b) **Malaya Archipelago**

(c) Indo-China (d) America

1811. 'Mangosteen' fruit contains seed. The development of mangosteen seed is from nuclear tissue hence propagates true to type through sexual propagation. This process of seed formation is called:

(a) Polyembryony (b) **Parthenogenesis**

(c) Apomixes (d) Parthenocarpy

1812. Mangosteen, (*Garcinia mangostina*) flower is sessile and botanically its fruit is:

(a) Berry (b) Drupe

(c) Pome (d) Nut

1813. *Manilkara hexandra*, 'Khirni' is rootstock most commonly used in:

(a) *Litchi chinensis* (b) *Cydonia oblonga*

(c) *Malus domestica* **(d) *Achras sapota***

1814. Mariana an important rootstock of plum has developed as a result of cross between:

(a) *Prunus munsoniana* and *P. cerasifera*

(b) *A. salicina* and *P. domestica*

(c) *P. salicina* and *P. simoni*

(d) *P. domestica* and *P. cerasifera*

1815. Marmalade is prepared from the fruits of:

(a) Citrus (b) Apple

(c) Mango (d) Guava

1816. Marmalades must have:

(a) 2% pectin (b) 1.25% pectin

(c) 0.75% pectin **(d) 1.0% pectin**

1817. Marmelosin an active ingredient in *Agle marmelos* is extracted from:

(a) Seed (b) Leaves

(c) Bark (d) AH the parts

1818. Mass selection is essentially a:

(a) Hybridization technique **(b) Population improvement**

(c) Mutation method (d) Clonal selection

1819. Mauritius is variety of;

(a) Grape **(b) Pineapple**

(c) Lime (d) None of these

1820. Mauritius which is a popular variety of pineapple is classified as?

(a) Late season (b) Early season

(c) Fruit throughout year **(d) Mid season**

1821. Maximum area and production of pineapple is in:

(a) W. Bengal (b) Assam

(c) Meghalaya (d) Manipur

1822. Maximum area of pomegranate is under

(a) Maharashtra (b) Punjab

(c) Karnataka (d) Assam

1823. Maximum number of fruit plants can be planted in the orchard by the system:

(a) Diagonal (b) Square

(c) Hexagonal (d) Rectangular

1824. Mclntosh is a variety of:

(a) Apple (b) Plum

(c) Peach (d) Pear

1825. Md. Kidwai award for biennium 1993-95 was conferred on which of the following horticulturists?

(a) Dr. I. S. Yadav

(b) Dr. Sant Ram and Dr. S.S Negi

(c) Dr. K.V Peter

(d) Dr. P Rethinam

1826. Meiotic cell division is called reduction division because:

(a) **Chro. number in the nuclei of daughter cells is reduced to half**

(b) It forms the basis for sexual reproduction

(c) Both of these

(d) None of these

1827. Mention form of sugar which is absent in *Vitis vinifera* at the time of ripening?

(a) **Sucrose** (b) Fructose

(c) Dextrose (d) Levulose

1828. Mention fruits which are indigenous to India?

(a) Mango, banana (b) Bael, jackfruit

(c) Aonla, Phalsa (d) **All of these**

1829. Mention the citrus species which is indigenotiito India?

(a) ***Citrus macroptera*** (b) *C. sinensis*

(c) *C. grandis* (d) *C. reticulata*

1830. Mention the disease caused by the Deficiency of Vit. B?

(a) Scurvy (b) **Beri-Beri**

(c) Pelagra (d) Night blindness

1831. Mention the family of Rambutan (*Nephelium lappaceum*) originated from Malayan Archipelago?

(a) Anacardiaceae (b) **Sapindaceae**

(c) Rosaceae (d) Sapotaceae

1832. Mention the fruit crop for which 'Colt' is a rootstock?

(a) **Cherry** (b) Pear

(c) Plum (d) Apple

1833. Mention the fruit crop in which metaxenia is commonly observed?

(a) **Date** (b) Grape

(c) None of these (d) Both of these

1834. Mention the fruit crop of which 'Chaubattia Alankar' and 'Chaubattia Madhu' are improved cultivars?

(a) Peach (b) Apple

(c) Apricot (d) Pear

1835. Mention the group of growth regulators which turn tendrils in shoot apices of grapevines into fruitful inflorescence?

(a) GA_3 (b) Ethylene

(c) Cytokinin (d) ABA

1836. Mention the group to which 'Maayan' Tropical Beauty', Winter Banana', Vered' 'Michael' 'Schomit' varieties of apple belongs to:

(a) Low chilling apple (b) High chilling apple

(c) Apple Scab resistant (d) None of the above

1837. Mention the largest fruit bearing amongst *Annona* species?

(a) *Annona muricata* (b) *A. glabra*

(c) *A. cherimoya* (d) *A. reticulata*

1838. Mention the name of apple hybrid developed as a result of a cross between Ambri x Golden Delicious?

(a) Suneheri (b) Lal ambri

(c) Rich-e-red (d) Maharajaji

1839. Mention the nutrient which improves fruit quality and gives rigidity to twigs, shoots, spurs, *etc.*?

(a) Calcium **(b) Potassium**

(c) Nitrogen (d) Phosphorus

1840. Mention the region of origin of cashew?

(a) Brazil (b) India

(c) Caspian Sea Area (d) South East Asia

1841. Mention the salt resistant grape varieties among the following pairs of grape varieties:

(a) Alamwick and Beauty Seedless

(b) Perlette and Khalili

(c) Anab-e Shahi, Bangalore Blue

(d) None of the above

1842. Mention the species to which 'Concord' 'Isabella' and 'Rogers' varieties of grape belong to?

(a) *Vitis vinifera* (b) *Muscadinia rotundifolia*

(c) *C. labrusca* (d) *V. rupestris*

1843. Chlorosis of midrib is due to the deficiency of

(a) Mn (b) Mg

(c) Zn **(d) Fe**

1844. Harvesting time of Subtropical pear is:

(a) Jan-Feb (b) March-April

(c) June-July (d) Sept- Octo

1845. Harvesting time of Strawberry is:

(a) May-June (b) Sept-Oct

(c) April-May **(d) Dec- Jan**

1846. Conc. Of SO_2 permitted for raisin preservation is:

(a) 250ppm **(b) 700ppm**

(c) 800ppm (d) 1000ppm

1847. Gynodioecious cross pollinated fruit crop is:

(a) Grape **(b) Fig**

(c) Banana (d) Pineapple

1848. Mallic acid is present in:

(a) Citrus (b) Guava

(c) Cherry (d) Pear

1849. Chromosome number of star fruit and star apple is:

(a) 24, 36 (b) 22, 34

(c) 28, 38 (d) 34, 48

1850. Bunchy top of banana was first observed in:

(a) Pune (b) Rome

(c) Fiji (d) None of these

1851. Bunchy top was got entry in India (1939, in Orissa) from:

(a) **Srilanka** (b) France

(c) Australia (d) China

1852. Recurrent type of apomixis is in:

(a) Citrus (b) Mango

(c) **Apple** (d) Mangosteen

1853. Rudimentary type of fruit is observed in which variety of sweet orange?

(a) Hamlin (b) Shamouti

(c) **Washington Novel** (d) Pineapple

1854. Vitamin -D is chemically is known as :

(a) Retinol (b) Cabalanin

(c) **Calciferol** (d) Tocopherol

1855. Lye peeling is done at a temperature of:

(a) 75⁰C (b) **84⁰C**

(c) 93⁰C (d) 105⁰C

1856. Daily requirement of Vitamin-A is:

(a) **1500 IU** (b) 1200 IU

(c) 750 IU (d) 2000 IU

1857. Fibril theory of jelly formation is given by:

(a) L. H. Meyer (b) E. C. Jhon

(c) **W. V. Cruess** (d) None of the above

1858. The point at which the dried product just become lumpy is known as:

(a) Danger point (b) Saturated point

(c) **Critical point** (d) Safety point

1859. First president of ICAR, New Delhi was

(a) Dr. B. P. Pal (b) **Habibullah**

(c) Rafi Ahmed Kidwai (d) Dr. Mangla Roy

1860. Number of State Agricultural Universities under ICAR are:

(a) 34 **(b) 38**

(c) 32 (d) 44

1861. ICAR Golden Jubilee Year was:

(a) 1979 (b) 1980

(c) 1988 (d) 1879

1862. Instrument was used for measurement of Leaf Area?

(a) Tendrometer **(b) Alanometer**

(c) Aplanometer (d) None

1863. Limiting factor for Cashew nut cultivation is:

(a) Temperature (b) Water

(c) Quality seedling (d) Harvesting

1864. Samrudhi variety belongs to which crop?

(a) Pineapple (b) Papaya

(c) Cocoa **(d) Arecanut**

1865. Filler crops are used in which type of planting system?

(a) Square (b) Hexagonal

(c) Quincunx (d) Rectangular

1866. Which is a good fungicide used for fumigating the soil?

(a) Formaldehyde (b) Vapam

(c) Sodium hypochloride **(d) All of these**

1867. Species of Apple is produced through apomixis:

(a) *Malus Sikkimensis* (b) *M. Hupensis*

(c) *M. Sergenti* **(d) All of these**

1868. In Temperate fruits, the "Root cutting" are prepared during the month of:

(a) January (b) February

(c) October **(d) December**

1869. Against phylloxera insect grape varieties are propagated during dormant season using

(a) Patch budding (b) **Chip budding**

(c) Ring budding (d) T-budding

1870. Ber powdery mildew is caused by:

(a) *Oidium erysiphoides* (b) *Uncinula necator*

(c) *Venturia inaequalis* (d) None of these

1871. Which variety of Banana belongs to Pomegranate:

(a) Gold Finger (b) Lady Finger

(c) Ney Poovan (d) All of these

1872. TSS of mango is:

(a) 20% (b) 25%

(c) 26% (d) 22%

1873. Rumani is used for dwarfing effect in:

(a) Dashehari (b) Langra

(c) Himsagar (d) Alphonso

1874. Vellaicolumban is used for dwarfing effect in:

(a) Dashehari (b) Langra

(c) Himsagar **(d) Alphonso**

1875. Number of perfect flower is highest in:

(a) Langra (b) Rumani

(c) Himsagar (d) All of these

1876. Insecticide act was passed :

(a) 1965 (b) 1966

(c) 1968 (d) 1962

1877. IIHR was established during:

(a) 1965 (b) 1966

(c) 1968 (d) 1962

1878. Any material which removes moisture without reacting with is known as:

(a) **Desiccant** (b) Dehydration

(c) Absorbent (d) Drier

1879. Which of the following is not ethylene inhibitor?

(a) AVG (b) MVG

(c) **NAA** (d) ABA

1880. Protein percent in coconut is:

(a) **4%** (b) 50%

(c) 10% (d) 8%

1881. Who discover enzyme?

(a) **Buchner** (b) Funk

(c) Kunain (d) Symner

1882. Volatile compound of orange is:

(a) **Citral** (b) UPA

(c) BHC (d) Naringin

1883. Which of the following group of yeast causing spoilage to fruit juices?

(a) Pichia (b) Trichosporon

(c) Rhodotorula (d) **Hanseniospora**

1884. At which pH range, the growth of bacteria takes place?

(a) 1.5-3.5 (b) **4.0-7.5**

(c) 8.0-10.5 (d) none

1885. The fruit whose colouring matter is insoluble in water is packed in which type of cans?

(a) SR cans (b) **Plain cans**

(c) Lacquered (d) Both (a) & (b)

1886. For canning of mango fruit, which types of cans are used?

(a) Plain (b) Lacquered

(c) SR cans **(d) both (a) & (b)**.

1887. Sticky or gummy jam is due to the:

(a) Low T.S.S **(b) High T.S.S**

(c) High pectin (d) Any other

1888. Jam contains ....acid?

(a) 0.5-0.6% (b) 1.5-2.5%

(c) 0.8-1.0% (d) 2.5-3.5%

1889. Acid content of jelly is:

(a) 0.3-0.4% **(b) 0.5-0.75%**

(c) 0.9-1.10% (d) 1.20-1.60%

1890. Corn syrup or glucose may be added along with cane sugar for avoiding?

(a) Sticky jam (b) **Crystallization of jam**

(c) Premature setting of jam (d) Avoiding shrinkage

1891. Which of the following fruit comes under rich in pectin and acid for jelly preparation?

(a) Jamun **(b) Ripe guava**

(c) Pear (d) Pineapple

1892. Use of immature fruits for preparation of jelly leads to

(a) Failure to set (b) **Foggy jelly**

(c) Formation of crystals (d) None of these

1893. At what ratio, the sweet orange and lime are taken for preparing the jelly?

(a) 1:1 **(b) 2:1**

(c) 3:1 (d) Either (a) or (b)

1894. What is the quantity of KMS for avoiding the browning of marmalade during storage?

(a) 0.090g (b) 0.11g

(c) 0.060g (d) 0.033g

1895. Fermentation of preserve and candy is due to:

(a) Low content of sugar

(b) Filling of preserve without cooling

(c) Inadequate blanching

(d) Cooking in heavy syrup

1896. Amchur is prepared from:

(a) Immature green mango fruits

(b) Mature green fruits

(c) Ripe mango fruits

(d) Both (b) & (c)

1897. Mango slab is also known as :

(a) Canned mango (b) Mango chutney

(c) Mango leather (d) Amchur

1898. What is conc. Of SO2 for adding in latex of papaya for extending the storage life of papain:

(a) 500ppm **(b) 750ppm**

(c) 1000ppm (d) 1500ppm

1899. Drying of latex is done at:

(a) 50-55°C (b) 75-80°C

(c) 40-45°C (d) 90-110°C

1900. What should be the conc. Of vinegar as a preservative?

(a) >2% **(b) <2%**

(c) 2-4% (d) None

1901. At 5-9⁰C temperature range, what is the storage life of mango?

(a) 2-3 weeks (b) 5-7 weeks

(c) 9-10 weeks (d) None

1902. At 10⁰C temperature range, what is the storage life of Banana?

(a) 1-2 weeks (b) 3-4 weeks

(c) 5-6 weeks (d) 7-8 weeks

1903. What is fruit juice percent of nector?

(a) 5% (b) 10%

(c) 20% (d) 15%

1904. What is fruit juice percent of cordial?

(a) 15% (b) 20%

(c) 25% (d) 30%

1905. What is fruit juice percent of natural juice?

(a) 55% (b) 75%

(c) 90% **(c) 100%**

1906. What is acidity percent of syrup?

(a) 1.3-1.5 (b) 1.0-1.2

(c) 1.7-1.9 (d) None

1907. Isopentyl acetate is volatile compound present in:

(a) Grape (b) Pineapple

(c) Banana (d) None of these

1908. Indicator plant of Xyloporosis is:

(a) Kagzi lime (b) Rangpur lime

(c) Sweet lime (d) Mandarin

1909. 'O' Horizon is absent in:

(a) Red soil (b) Organic soil

(c) Arable soil **(d) Sandy soil**

1910. Leading cultivar contributing 58% products:

(a) **Basrai** (b) Dwarf Cavendish

(c) Robusta (d) All of these

1911. After nut set of coconut, the nut is harvested in

(a) 5 months (b) 6 months

(c) 3 months **(d) 11 months**

1912. The ploidy level of European plum is

(a) Diploid (b) Triploid

(c) Hexaploid (d) Tetraploid

1913. First successful transgenic fruit plants are produced in

(a) Pear (b) Walnut

(c) Cherry (d) Apple

1914. *Tylencaulus semepenitrans* a nematode is a serious problem in

(a) Citrus (b) Guava

(c) Banana (d) Coconut

1915. Low chill peach variety is:

(a) Kanto-5 (b) Shimizo Hakuto

(c) Shan-e- Punjab (d) July Elbarta

1916. Fruit crop resistant to wet soil is:

(a) Peach (b) Apricot

(c) Quince **(d) Pear**

1917. Skiffing is done:

(a) August-September (b) June-July

(c) December-January (d) February-March

1918. Vengurla-3 is hybrid between:

(a) Ansur-1 x Vetore-56 **(b) Midnapur red x Vetore-56**

(c) Anur Early x Kottekar (d) None of these

1919. NCD stands for
(a) National Cashew Dwarf
(b) National Coco Development
(c) National Coconut Development
(d) Natural Cross Dwarf

1920. S. S. Green is a variety of
(a) Cashewnut **(b) Coconut**
(c) Tea (d) Coffee

1921. Cocoa canker is due to:
(a) Fungus (b) Bacteria
(c) MLO's (d) Virus

1922. Shelling percentage of cashewnut is:
(a) 13% (b) 23%
(c) 33% **(d) 43%**

1923. BPP-5 is a variety of cashew which is released through:
(a) Introduction (b) Hybridization
(c) Selection (d) Mutation

1924. White stem borer is major pest of :
(a) Coconut **(b) Coffee**
(c) Cashewnut (d) Cocoa

1925. CCRI is situated at (in case of coffee)
(a) Kassargod **(b) Balahanpur**
(c) Trissur (d) Cochin

1926. Which manufacturing stage decides the quality of tea?
(a) Withering (b) Rolling
(c) Fermentation (d) Drying

1927. In type of processing series of rollers are used in

(a) CTC (b) Orthodox

(c) Both (a) & (b) (d) None

1928. Rolling of manufacturing of tea is carried out for

(a) 20 min (b) 15 min

(c) 30 min (d) 55 min

1929. Coffee is a ....plant

(a) Long day **(b) Short day**

(c) Day neutral (d) Either (a) or (b)

1930. Genus coffee consists of about ....species

(a) 50 (b) 55

(c) 60 (d) 70

1931. *C. arabica* is:

(a) Diploid (b) Triploid

(c) Tetraploid (d) Hexaploid

1932. Which species of coffee is not a cross pollinated?

(a) *C. arabica* (b) *C. liberica*

(c) *C. canephora* (d) *C. robusta*

1933. The elevation range for Arabica coffee:

(a) 500-1000 m MSL (b) 1000-1500 m MSL

(c) 1600-2000 m MSL **(d) 2100-2500 m MSL**

1934. San Roman is a hybrid of:

(a) Coconut (b) Cashewnut

(c) Coffee (d) Tea

1935. Scuffling is done in:

(a) Cocoa **(b) Coffee**

(c) Tea (d) Oil Palm

1936. Harvesting of Cashewnut is:
(a) Dec-Jan (b) Sep- Oct
(c) June-July **(d) March-April**

1937. Cashew apple is a rich source of:
(a) Vit-A (b) Vit-B
(c) Vit-C (d) Protien

1938. Palm oil is extracted from:
(a) Mesocarp & Kernel (b) Only Kernel
(c) Only Mesocarp (d) None of these

1939. Stripping is done in:
(a) Coffee (b) Tea
(c) Cocoa **(d) Oil Palm**

1940. Per hectare yield of Oil Palm?
(a) 10 tonnes **(b) 4 tonnes**
(c) 2 tonnes (d) 15 tonnes

1941. Cal-17 is a variety of?
(a) Coconut (b) C. Nut
(c) Arecanut (d) Oil Palm

1942. NRC for oil palm is located is:
(a) Eluru in A. P. (b) Tiruchirapalli, T. N.
(c) Kottayam, Kerala (d) None of these

1943. In which group of oil palm, shell is present?
(a) Dura (b) Pisifera
(c) (a) & (b) (d) None of these

1944. In which group of oil palm, shell is absent?
(a) Dura **(b) Pisifera**
(c) (a) & (b) (d) None of these

1945. S_2d_2 taping system has ...... intensity?

(a) 100% (b) 67%

(c) 200% (d) 50%

1946. Which system of tapping is used for budded plants of rubber?

(a) S_2d_2 (b) S_2d_3

(c) S_2d_1 (d) Both (a) & (b)

1947. S_2d_1 taping system has ...... intensity?

(a) 100% (b) 67%

(c) 200% (d) 50%

1948. Which system of tapping of rubber is recommended for small growers?

(a) S_2d_2 (b) S_2d_3

(c) S_2d_1 (d) Both (a) & (b)

1949. Vascular streak is a serious disease of:

(a) Coffee (b) Tea

(c) Cocoa (d) Oil Palm

1950. ....is the second important commodity in the world trade after petroleum products

(a) Tea (b) Cocoa

(c) Coffee (d) Cashewnut

1951. Ablation is done in:

(a) Tea (b) Coffee

(c) Oil Palm (d) Cocoa

1952. SAS -1 is a cultivar of:

(a) Tea (b) Cocoa

(c) Arecanut **(d) Oil Palm**

1953. Cocoa seed remains viable for:

(a) 9 days **(b) 7 days**

(c) 15 days (d) 23 days

1954. How many hours of sunshine is required for profuse growth and production to the coconut palm?

(a) 500 (b) 1000

(c) 700 **(d) 2000**

1955. Mohitnagar is a variety of :

(a) Arecanut (b) Coconut

(c) Coffee (d) Date Palm

1956. Mohitnagar is a variety of Arecanut recommended for which state:

(a) MH **(b) W.Bengal**

(c) Kerala & HR (d) Both (a) & (b)

1957. Which of the following country is the largest producer of fruits?

(a) Brazil (b) India

(c) China (d) U.S.A.

1958. What is the place of India in world in fruit production?

(a) First **(b) Second**

(c) Third (d) Fourth

1959. Fruit which contains maximum calories is

(a) Date palm (b) Karonda

(c) Custard apple (d) Aonla

1960. 'Chhuhara' is an important variety of

(a) Date palm (b) Grape

(c) Guava **(d) Ber**

1961. Which of the following fruit tree is insensitive to *acidity* and salinity?

(a) Phalsa **(b) Aonla**

(c) Guava (d) Plum

1962. The percentage of Vitamin C is maximum in

(a) Grape fruit (b) Pummelo

(c) Lemon (d) Orange

1963. The ideal intercrop in citrus orchard is

(a) Bajra (b) Chillies

(c) Peas (d) Cluster bean

1964. Decrease in the rate of respiration in plants will

(a) Decrease water absorption (b) Increase water absorption

(c) Decrease transpiration (d) Stop water absorption

1965. Chlorosis in plants is caused due to deficiency of

(a) Manganese and Sulphur (b) Boron and Phosphorus

(c) Calcium and Oxygen **(d) Nitrogen and Magnesium**

1966. Fruit formation is suppressed due to deficiency of

(a) Sulphur **(b) Boron**

(c) Calcium (d) Magnesium

1967. Which of the following fruit crop is adversely affected by mealy bug?

(a) Papaya **(b) Mango**

(c) Banana (d) Guava

1968. ICAR institute located in Kasargod is :

(a) CPRI **(b) CPCRI**

(c) CARI (d) CRRI

1969. Which one of the following country is the largest producer of tea in the world?

(a) China **(b) India**

(c) Indonesia (d) Sri Lanka

1970. Which of the following family, the 'Loquat' belongs to?

(a) Rosaceae (b) Juglandaceae

(c) Anacardiaceae (d) Apocynaceae

1971. At what concentration level, the sugar acts as preservative?

(a) 66% (b) 60%

(c) 50% (d) 40%

1972. Multistoryed cropping is popular in

(a) Cashewnut Plantation **(b) Coconut Plantation**

(c) Cocoa Plantation (d) Passion Fruit Plantation

1973. 'Sonaka' is a variety of

(a) Pomegranate (b) Sapota

(c) Grape (d) Sweet Cherry

1974. Central Institute of Arid Horticulture is located at

(a) Jodhpur **(b) Bikanr**

(c) Jaipur (d) Udaipur

1975. 'Halawy' is a popular variety of

(a) Coconut **(b) Date palm**

(c) Cashewnut (d) Arecanut

1976. In which state of India, fig is grown?

(a) MP (b) Uttar Pradesh

(c) Kerala **(d) Gujarat**

1977. Guava is a native of

(a) South-East Asia (b) Tropical Africa

(c) Tropical America (d) Sub-tropical India

1978. 'Pusa Urvashi' is a variety of

(a) Mango (b) Guava

(c) Grape (d) Jackfruit

1979. Dioecious form of sex is found in

(a) Aonla (b) Citrus

(c) Litchi **(d) Papaya**

1980. Which of the following climate is suitable for Banana?

(a) Cold (b) Dry

(c) Humid (d) Hot

1981. The National Fruit of India is
(a) Apple (b) Banana
(c) Mango (d) Guava

1982. The edible part of Litchi is -
(a) Placenta (b) Endosperm
(c) Mesocarp **(d) Aril**

1983. Which of the following state is largest producer of cashewnut?
(a) Karnataka **(b) Kerala**
(c) Tamil Nadu (d) MH

1984. A mature and ripened ovary containing seed is referred as
(a) Stone (b) Embryo
(c) Fruit (d) None of these

1985. Which of the following fruit is known as "bathroom fruit"?
(a) Banana (b) Jackfruit
(c) Grapefruit **(d) Mango**

1986. Which of the following plant hormone is used for berry elongation in grape
(a) IAA (b) ABA
(c) GA_3 (d) NAA

1987. Horticultural Society of India was founded in
(a) 1955 **(b) 1942**
(c) 1952 (d) 1932

1988. National Botanical Research Institute is located at
(a) Lucknow (b) Bangalore
(c) New Delhi (d) None of these

1989. Which of the following is origin centre of Avocado?
(a) Tropical America **(b) Central America**
(c) South-East Asia (d) China

1990. Which of the following is not propagated through seed?

(a) Phalsa (b) Mangosteen

(c) Loquat (d) Papaya

1991. On the basis of respiration rate, grapes are comes under class of -

(a) Very low **(b) Low**

(c) Moderate (d) High

1992. Mulberry fruit is botanically known as

(a) Blausta **(b) Sorosis**

(c) Amphisarca (d) Syconium

1993. Which of the following fruit is richest source of Riboflavin?

***(a)* Bael** (b) Cashewnut

(c) Papaya (d) Pineapple

1994. Which of the following fruit is highest source of carbohydrate (%)?

(a) Date palm (b) Banana

(c) Apricot (d) Cashewnut

1995. Richest source of ascorbic acid is

(a) Aonla (b) Guava

(c) Barbados Cherry (d) Lemon

1996. Which of the following fruit contain highest fat?

(a) Avocado (b) Almond

(c) Cashewnut **(d) Walnut**

1997. 'Alphonso', a variety of mango is mostly grown in

(a) Punjab (b) UP

(c) **Maharashtra** (d) Bihar

1998. In aonla, flowering occurs during the month of

(a) January (b) **March**

(c) May (d) July

1999. Which of the following rootstock of citrus is/are tolerant to tristeza virus?

(a) Rangpur lime (b) Rough lemon

(c) Citrange **(d) All of these**

2000. Which of the Apple rootstock is/are resistant to wooly aphid?

(a) M9 (b) M27

(c) Both (d) None of these

2001. Banana fruit is botanically known as

(a) Pome (b) Drupe

(c) **Berry** (d) Stone

2002. Ploidy level of Aonla is

(a) Auto-tetraploid (b) Auto-hexaploid

(c) Auto-octaploid (d) None of these

2003. Seedlessness in banana is due to

(a) Mutation (b) Lack of ovules

(c) Vegetative parthenocarpy (d) All of the above

2004. Botanically Guava fruit is

(a) Pome (b) Stone

(c) Berry (d) Drupe

2005. *Fragaria vesca* is commonly known as -

(a) Raspberry (b) Cranberry

(c) Strawberry (d) Blueberry

2006. Redness in apple is due to presence of

(a) Anthocyanin (b) Carotene

(c) Lycopene (d) Xanthophyll

2007. The best method of irrigation for mango planted in sandy soil is

(a) Sprinkling **(b) Drip**

(c) Channel (d) Bed

2008. Latex of which plant is used for making chewing gum?

(a) Ficus (b) **Sapota**

(c) Rubber (d) None of these

2009. Which of the following fruit is used in preparation of the famous drink Feni' ?

(a) **Cashewnut** (b) Custard Apple

(c) Cashew Apple (d) Raspberry

2010. Mango variety which is suitable for high density planting –

(a) Mallika (b) **Amrapalli**

(c) Neelum (d) Sindhu

2011. Bunchy top of banana is caused by

(a) **Virus** (b) Fungi

(c) Mycoplasma (d) Bacteria

2012. The stem above the ground in banana is known as

(a) Rhizome (b) **Pseudostem**

(c) Sucker (d) Tuber

2013. Botanically Kinnow is a

(a) Citrange (b) **Mandarin**

(c) Tangerine (d) None of these

2014. 'Roopa' is an improved variety of

(a) Guava (b) Litchi

(c) Pomegranate (d) **Walnut**

2015. Which of the following fruit is known as Hesperidium ?

(a) Custard apple (b) Apple

(b) **Orange** (d) Litchi

2016. Which of the following is not a mango variety?

(a) Arka Anmol (b) Arka Aruna

(c) **Arka Hans** (d) Arka Puneet

2017. Which of the following fruit is not indigenous to India?

(a) Mango (b) Jackfruit

(c) Aonla **(d) Pineapple**

2018. In Ber, pruning is done during

(a) Rainy Season (b) Winter

(c) Summer **(d) Spring**

2019. Which of the following fruit is richest source of iron?

(a) Apple (b) Datepalm

(c) Karonda (d) Mango

2020. Ploidy level of edible banana is

(a) Diploid **(b) Triploid**

(c) Tetraploid (d) None of these

2021. Which of the following is a polyembryonic variety of mango?

(a) Chandrakaran (b) Salem

(c) Olour **(d) Alphonso**

2022. At what temperature fruits are blanched before canning

(a) 80° C (b) 90° C

(c) 100° C (d) 105° C

2023. Which varieties of mango have regular bearing habit?

(a) Langra (b) Alphonso

(c) Amrapali (d) Dasheri

2024. Gynodioceious cultivar of papaya

(a) Pusa Giant **(b) Pusa Majesty**

(c) Pusa Dwarf (d) None of these

2025. Botanically pineapple fruit is known as

(a) Berry (b) Pome

(c) Sorosis (d) Balusta

2026. Which of the following fruit in India has highest productivity?

(a) Citrus **(b) Banana**

(c) Grapes (d) Mango

2027. Mango seed lose their viability within the

(a) 2-3 weeks **(b) 4-5 weeks**

(c) 6-8 weeks (d) 9-10 weeks

2028. Which of the following fruit tree is heavily pruned?

(a) Mango (b) Aonla

(c) Ber (d) Guava

2029. Protandry is associated with

(a) Banana (b) Grape

(c) Papaya **(d) Walnut**

2030. Which of the following fruit crop is most suceptible to water logging?

(a) Ber (b) Banana

(c) Papaya (d) Pineapple

2031. Sweetest sugar present in fruit

(a) Glucose (b) Galactose

(c) Fructose (d) Sucrose

2032. Karonda belongs to family

(a) Apocynaceae (b) Boraginaceae

(c) Euphorbiaceae (d) Myrtaceae

2033. Botanically Ber fruit is known as

(a) Berry (b) Pome

(c) Drupe (d) Stone

2034. Litchi is commercially propagated through

(a) Runner (b) Grafting

(c) **Air Layering** (d) Budding

2035. Highly shape tolerant fruit crop is -

(a) Kiwi fruit (b) Rambutan

(c) Carambola (d) Durian

2036. Which of the following fruit is the richest source of protein?

(a) Banana (b) Almond

(c) Cashewnut (d) Walnut

2037. Botanically tamarind fruit is known as

(a) Berry **(b) Pod**

(c) Pome (d) Stone

2038. In North India, Grapes are pruned in the month of

(a) January (b) March

(c) July (d) October

2039. Papaya is a rich source of

(a) Vitamin A (b) Vitamin B

(c) Vitamin C (d) Vitamin E

2040. Which of the following fruit crop is most suitable for filler crop in mango orchard?

(a) Aonla (b) Phalsa

(c) Papaya (d) Grape

2041. International Institute of Horticulture is situated at

(*a*) Brazil (b) China

(c) USA (d) Italy

2042. Fruit which has earliest bearing after planting

(a) Banana (b) Pineapple

(c) Papaya **(d Strawberry**

2043. Pineapple belongs to the family of -

(a) Anacardiaceae **(b) Bromeliaceae**

(c) Juglandaceae (d) None of these

2044. Which of the following fruit is most suitable for marmalade preparation?

(a) Guava (b) Litchi

(c) Orange (d) Apple

2045. The inception of ripening in grape is termed as

(a) Breakeven point (b) Colour break

(c) Turning stage **(d) Verasion**

2046. Which of the following fruit crop pollination is done by hand ?

(a) Grape **(b) Date palm**

(c) Mango (d) Pineapple

2047. Litchi is native of

(a) North China (b) North America

(c) India **(d) South China**

2048. 'Sindhu', a variety of mango is a cross between

(a) Alphonso x Amrapali **(b) Ratna x Alphonso**

(c) Ratna x Neelum (d) None of these

2049. Which chemical serves as an alternative to caprification in fig.

(a) SADH **(b) 1-CPA**

(c) BOA (d) CCC

2050. Seedlessness in lemon is due to

(a) Ovule Sterility (b) Parthenocarpy

(c) Self-incompatibihty (d) Parthenogenesis

2051. Which of the following citrus fruit is monoembryonic ?

(a) Mandarin (b) Grape fruit

(c) Pummelo (d) Citron

2052. Late ripening variety of Ber -

(a) Gola (b) Kaithi

(c) Umran (d) All of the above

2053. Witches broom is the major disorder of which crop?

(a) Coffee (b) Tea

(c) Cocoa (d) Betelvine

2054. In hexagonal planting system how much more plants can accomodate than the square system

(a) 10% **(b) 15%**

(c) 25% (d) 40%

2055. Father of systematic pomology

(a) Darwin (b) Mendel

(c) Decandole (d) None of these

2056. Plant growth regulator used for fruit thining

(a) IAA (b) NAA

(c) GA_3 (d) C_2H_4

2057. Which growth regulator is used to reduce period of chilling requirement?

(a) Cytokinin (b) IBA

(c) NAA **(d) GA_3**

2058. Which of the following is an example of aggregate fruit?

(a) Guava (b) Citrus

(c) Pineapple **(d) Strawberry**

2059. Dwarf mutant cultivar of papaya is

(a) Pusa Dwarf **(b) Pusa Nanha**

(c) Pusa Majesty (d) Pusa Giant

2060. Cashewnut is originated from

(a) India (b) Soufh-East Asia

(c) Brazil (d) China

2061. The yellow colour pigment in papaya fruit is due to

(a) Anthocyanin **(b) Caricaxanthin**

(c) Carotene (d) Xanthophyll

2062. Type of parthenocarpy in 'Sindhu' variety of mango is

(a) Stimulative (b) Vegetative

(c) Stenospermocarpy (d) None of these

2063. Which of the following fruit crop bear fruits twice in a year

(a) Jackfruit (b) Pomegranate

(c) Guava **(d) All of the above**

2064. Which of the following is associated with bael?

(a) Annonine (b) Ficin

(c) Lycopene **(d) Marmelosin**

2065. Parthenocarpic variety of a grape is -

(a) Perlette (b) Arka Hans

(c) Black Corianth (d) None of these

2066. Mango hybridization work was started in the year :

(a) 1901 **(b) 1911**

(c) 1921 (d) 1931

2067. Planting material of pineapple is

(a) Sucker **(b) Slip**

(c) Crown (d) Cutting

2068. Pomegranate variety 'Jalore Seedless' is recommended for cultivation in

(a) Gujarat (b) UP

(c) Maharashtra **(d) Rajasthan**

2069. Salt resistant rootstock of grape -

(a) Salt Creek (b) Dogridge

(c) both (a) and (b) (d) None of these

2070. Bronzing in guava is caused due to deficiency of

(a) Copper (b) Manganese

(c) Phosphorus (d) Magnesium

2071. Which plantation crop is heliotropic in nature?

(a) Avecanut (b) Coffee

(c) Coffee **(d) Coconut**

2072. Which of the following fruit(s) is/are monocotyledonous?

(a) Banana (b) Pineapple

(c) Date palm **(d) All of the above**

2073. Plant growth regulator used for rooting in cuttings

(a) ABA (b) NAA

(c) IBA (d) GA3

2074. Which of the following fruit follow double sigmoid curve?

(a) Fig (b) Ber

(c) Mango (d) Pear

2075. Fruits which yellow in colour are rich source of

(a) Vitamin A (b) Vitamin B

(c) Vitamin C (d) Vitamin E

2076. Best rootstock for citrus in South India is

(a) Citrangc **(b) Rangpur lime**

(c) Mandarin (d) All of the above

2077. Banana inflorescence is known as

(a) Catkin (b) Raceme

(b) Spadix (d) Umbel

2078. Papaya is commercially propagated through

(a) Budding (b) Cuttings

(c) Layering **(d) Seed**

2079. Pollen sterility is observed in -

(a) Aonla **(b) Ber**

(c) Grape (d) Pomegranate

2080. Metaxenia is commonly observed in
(a) Apple (b) **Date palm**
(c) Grape (d) Mango

2081. 'Cricket Ball' is a popular variety of
(a) Mango (b) Apple
(c) **Sapota** (d) Custard Apple

2082. Almond is commercially propagated through -
(a) Layering (b) Cutting
(c) **Tongue grafting** (d) Stone grafting

2083. Which of the following state produce maximum Lime in India?
(a) Rajasthan (b) Uttar Pradesh
(c) **Andhra Pradesh** (d) Maharashtra

2084. Which of the following is a popular cultivar of a ber tolerant against salinity and alkalinity
(a) Umran (b) **Gola**
(c) Dodia (d) All of the above

2085. Cashew is propagated through -
(a) Air layering (b) **Epicotyl grafting**
(c) Stooling (d) Cuttings

2086. Which of the following is/are self-incompatible cultivar(s) of mango?
(a) Chausa (b) Bombay green
(c) Langra (d) **All of the above**

2087. Native place of Date palm-
(a) India (b) Afghanistan
(c) **Iraq** (d) Indonesia

2088. Fruit cracking in pear is due to deficiency of
(a) B (b) **Zn**
(c) Mg (d) Cu

2089. 'Pusa Early Dwarf is a popular cultivar of -

(a) Papaya (b) **Strawberry**

(c) Pear (d) Guava

2090. Which of the following is a cold tolerant species of papaya ?

(a) *Carica papaya* (b) *Carica cauliflora*

(c) *Carica candicona* ***(d) Carica candamarcensis***

2091. Botanically, sapota fruit is known as -

(a) Pome **(b) Drupe**

(c) Berry (d) Blausta

2092. 'Arka Mridula' is a variety of -

(a) Guava (b) Mango

(c) Pomegranate (d) Sapota

2093. 'Tas-e-Ganesh' is a popular variety of which fruit crop?

(a) Mango (b) Guava

(c) Grape (d) Pomegranate

2094. Because of Polyembryony which of the following fruit can be grown by seed?

(a) Lemon **(b) Lime**

(c) Kinnow (d) Malta

2095. 139. Amphimixis is also known as

(a) Asexual method **(b) Sexual method**

(c) Vegetative method (d) None of these

2096. In Guava fruits which bahar is recommended?

(a) Ambe bahar (b) Hasth bahar

(c) Mrig bahar (d) All of the above

2097. Stooling is common method of propagation in

(a) Aonla **(b) Guava**

(c) Mango (d) Pomegranate

2098. Self-incompatibility does not exist in which mango variety?

(a) Chausa (b) Langra

(c) Dashehari **(d) Mallika**

2099. In hilly areas fruit plants should be planted in -

(a) Rectangular system (b) Square system

(c) Quincunx system **(d) Contour system**

2100. Which of the following is not a climacteric fruit?

(a) Banana **(b) Grape**

(c) Mango (d) Sapota

2101. Mango variety which is resistant to malformation -

(a) Alphonso (b) Amrapali

(c) Bhadauran (d) Neelum

2102. The elements which improve the quality of fruits is

(a) Boron (b) Calcium

(c) Potassium (d) Zinc

2103. Kinnow is a hybrid of :

(a) Sweet orange (b) Sour orange

(c) Mandarin (d) Poncirus

2104. Origin of Apple is from :

(a) South America (b) North America

(c) S.W. Asia (d) S.E. Asia

2105. Mango variety which is free from spongy tissue is :

(a) Alphonso (b) Manjira

(c) Ratna (d) Rumani

2106. For regular bearing in mango which growth regulator is recommended?

(a) GA3 (b) IAA

(c) NAA **(d) Paclobutrazol**

2107. The ratio of retention of male plants in papaya plantation should be :

(a) 7 : 3 (b) 8 : 2

(c) 9:1 (d) 19:1

2108. Parthenocarpic fruits are :

(a) False fruits (b) Composite fruits

(c) Seedless fruits (d) None of these

2109. Which of the following is not a tropical fruit?

(a) Sapota (b) Jackfruit

(c) Pineapple **(d) Apricot**

2110. Which of the following is a deciduous fruit tree?

(a) Mango (b) Citrus

(c) Pear (d) Litchi

2111. Botanically, Fruit of Fig is known as :

(a) Capsule (b) Sorosis

(c) Drupe **(d) Cyconus**

2112. Edible portion of grape is :

(a) Juicy placental hairs **(b) Pericarp and placenta**

(c) Fleshy layer of pericarp (d) Pericarp and thalamus

2113. Titron, Kelsey and Alpha are the popular variety of :

(a) Peach **(b) Plum**

(c) Pear (d) Phalsa

2114. 'Red Sitaphal' is a variety of :

(a) Guava (b) Pineapple

(c) Sapota **(d) Custard apple**

2115. Parthenogenesis is associated with which of the following fruit?

(a) Banana (b) Grape

(c) Mangosteen (d) Jamun

2116. Which of the following peach variety is suitable for Northern Plains?

(a) Alton (b) Quetta

(c) J.H.Hale (d) Prabhat

2117. Which of the following is standard spacing for planting of Ber?

(a) 6 x 6 m **(b) 8x8m**

(c) 10 x 10 m (d) 4 x 4 m

2118. Which of the following is a 'Off season variety of Mango'?

(a) Fazli (b) Mallika

(c) Niranjan (d) Olour

2119. The term 'three quarterful or full three quarter' is used to denote fruit maturity in which of the following fruit crop?

(a) Mango (b) Papaya

(c) Pineapple **(d) Banana**

2120. Generally below which temperature chilling injury occurs in mango:

(a) 8°C **(b) 10 °C**

(c) 12 °C (d) 14 °C

2121. 'Apple Colour' is a popular variety of :

(a) Apple (b) Peach

(c) Pear **(d) Guava**

2122. Hard wood stem cutting is commercial method of propagation in:

(a) Mango (b) Aonla

(c) Grape (d) Guava

2123. In Apples, pollination is done by :

(a) Air (b) Birds

(c) Insects (d) None of these

2124. Best planting material in Banana is :

(a) Water suckers (b) Rhizome

(c) Sword Suckers (d) Bulb

2125. 'T' budding is also referred as :

(a) Patch budding (b) Ring budding

(c) Shield budding (d) Annular budding

2126. Sunset is a variety of :

(a) Mango (b) Grape

(c) Litchi **(d) Papaya**

2127. Which of the following is a post harvest physiological disorder of Mango?

(a) Black tip (b) Internal necrosis

(c) Sap burn (d) None of these

2128. Mango is commercially propagated through :

(a) Inarching (b) Layering

(c) Cutting (d) Veneer grafting

2129. Which of the following country is largest producer of banana?

(a) China (b) Brazil

(c) India (d) Belgium

2130. Ploidy level of cultivated strawberry -

(a) Diploid (b) Triploid

(c) Tetraploid **(d) Octaploid**

2131. Which of the following is a common pollinator of Mango?

(a) Bumble bee **(b) House fly**

(c) Wasp (d) None of these

2132. Which of the following country is largest producer of Pineapple in the world ?

(a) Philippines (b) China

(c) Thailand (d) Brazil

2133. Largest producer of Papaya in India is :

(a) West Bengal (b) Uttar Pradesh

(c) Karnataka (d) Maharashtra

2134. Which of the following fruit comes under *Anacardiaceae* family ?

(a) Walnut **(b) Cashewnut**

(c) Jamun (d) Arecanut

2135. Plant parts that first shows the symptoms of Magnesium deficiency:

(a) New leaves (b) Leaf tip

(c) Old leaves (d) Middle leaves

2136. Which of the following country is largest producer of Cardamom in the world ?

(a) Argentina (b) China

(c) India (d) Brazil

2137. 'Arka Shyam' is a popular hybrid of -

(a) Jamun (b) Mango

(c) Grape (d) Guava

2138. 'Sigatoka' is associated with -

(a) Mango **(b) Banana**

(c) Apple (d) Pineapple

2139. 'Foster' is a popular variety of :

(a) Grape (b) Apple

(c) Peach **(d) Grape fruit**

2140. Guava fruits of 'Mrig bahar' are ripe during :

(a) July-September **(b) November-January**

(c) February-April (d) February-March

2141. Which of the following is/are correct with Sapota ?

(a) Slow growing evergreen tree

(b) Flowering and fruiting are terminal in the axils of leaves

(c) Requires high temperature and humidity

(d) All of the above

2142. Pomegranate can successfully propagated through :

(a) Hardwood cutting (b) **Air layering**

(c) Root suckers (d) All of the above

2143. Which of the following is/are essential for successful cultivation of Litchi

(a) High humidity (b) Rich deep soil

(c) Abundant moisture **(d) All of the above**

2144. 'Propping' an intercultural operation is followed in :

(a) Mango (b) Apple

(c) Banana (d) Guava

2145. 'Bishop' is a popular variety of :

(a) Pineapple (b) Papaya

(c) Sweet orange (d) Lemon

2146. Which of the following is a variety of Pummelo?

(a) Duncan (b) Villafranca

(c) Hamlin **(d) Kao Pan**

2147. International Society for Horticultural Science (ISHS) has its headquarters at :

(a) Italy **(b) Belgium**

(c) Germany (d) Brazil

2148. Date palm variety which can eaten as raw :

(a) Barhee **(b) Halawy**

(c) Both (a) and (b) (d) None

2149. Coconut Development Board comes under :

(a) Ministry of Commerce

(b) Ministry of Agriculture

(c) Ministry of Fruit Processing and Post harvest management

(d) ICAR

2150. Which fruit is called 'Miracle fruit of China' :

(a) **Kiwi fruit** (b) Mulberry

(c) Cherry (d) Litchi

2151. National Horticulture Board (NHB) was established in :

(a) 1980 (b) 1955

(c) 1952 (d) **1984**

2152. Which fruit is referred as strictly sub-tropical crop owing to its exact climatic requirement:

(a) Mango (b) **Loquat**

(c) Litchi (d) Lemon

2153. Macadamia nut *(Macadamia ternifolia)* belongs to family:

(a) **Proteaceae** (b) Juglandaceae

(c) Apocynaceae (d) None of these

2154. Macadamia nut was introduced in India from :

(a) Netherlands (b) Italy

(c) Germany (d) **Australia**

2155. 'Furete' is an improved variety of :

(a) Almond (b) Walnut

(c) **Avocado** (d) Olive

2156. The term of *Parthenocarpy* was coined by :

(a) Olmo, H.P. (b) Thompson

(c) **Stout** (d) Negls.S.

2157. 'Calyptra' is a term associated with which of the following fruit crops?

(a) **Grape** (b) Litchi

(c) Cherry (d) Phalsa

2158. Abortion of embryo before maturation which results in seedlessness is a problem in improvement of :

(a) Sapota (b) **Grape**

(c) Guava (d) Banana

2159. 'Laksha Ganga' is an improved hybrid variety of :

(a) Cashewnut **(b) Coconut**

(c) Arecanut (d) Cocoa

2160. The maximum papain yielding papaya variety is :

(a) CO-5 (b) CO-2

(c) Coorg Honey Dew (d) None of these

2161. Polyembryony in coconut was reported into :

(a) Lacadive (b) T x D

(c) Gangabondam (d) Aras kere Tall

2162. Dollar earning fruit crop in India is :

(a) Mango (b) Banana

(c) **Cashew nut** (d) Arecanut

2163. Barbados cherry *(Malphigra puncifolia* or M. *glabra)* belongs to family:

(a) Sterculiaceae (b) Guttifereae

(c) Apocynaceae **(d) Malphigiaceae**

2164. Durian (Civet fruit) *[Durian zibenthinus]* belongs to family:

(a) Bombaceae (b) Sterculiaceae

(c) Guttifereae (d) Rosaceae

2165. Grape was introduced in India from :

(a) Iran (b) Pakistan

(c) **Afghanistan** (d) Germany

2166. Jamun *(Syzigium cumini)* is also called as :

(a) Natal plum **(b) Java plum**

(c) Rose plum (d) None of these

2167. Central Institute of Sub-tropical Horticulture (CISH) was initially:

(a) Central Mango Research Station (CMRS)

(b) Central Mango Research Institute (CMRI)

(c) Mango Research Institute of India (MRII)

(d) National Centre on Mango (NCM)

2168. 'Mauritius' a popular variety of pineapple is classified as :

(a) Early season (b) **Mid season**

(c) Late season (d) Fruit throughout year

2169. 'Theobroma' in cocoa is extracted from :

(a) Bark (b) Seed

(c) Fruit Peel (d) Root

2170. 'Old chicks' is the variety of :

(a) Tea **(b) Coffee**

(c) Cocoa (d) Cashew

2171. Which beverage crop is referred as "Food of God"?

(a) Tea (b) Coconut

(c) Cocoa (d) Coffee

2172. Two leaf and a bud is type of plucking in :

(a) Tea (b) Coffee

(c) Cocoa (d) None of these

2173. Wooly apple aphid a serious pest of apple got its entry into India from :

(a) Pakistan (b) China

(c) France **(d) England**

2174. Cocoa belongs to family :

(a) Sterculiaceae (b) Bromileaceae

(c) Anonaceae (d) Palmae

2175. 'Amrit sagar' is a variety of :

(a) Ber (b) Guava

(c) Banana (d) Papaya

2176. Growth curve of fruit development in apple :

(a) Triple sigmoid **(b) Single sigmoid**

(c) Double sigmoid (d) No growth curve followed

2177. 'Mahal' *(Pyrus pashia)* is a rootstock of :

(a) Almond (b) Cherry

(c) Plum **(d) Pear**

2178. Fruit crop suitable for cold desert and cold arid zone is :

(a) Cranberry (b) Blueberry

(c) Avocado **(d) Apricot**

2179. Chinese date is common name of :

(a) Ber (b) Aonla

(c) Litchi (d) Phalsa

2180. Origin of Mandarin Orange is :

(a) West Indies **(b) China**

(c) India (d) Thailand

2181. Variety of papaya producing only female plants :

(a) Pusa Dwarf **(b) Sunrise Solo**

(c) Pusa Delicious (d) Pusa Giant

2182. In which fruit crop, Pond apple *(Annona glabra)* is used as a rootstock?

(a) Custard apple (b) Jamun

(c) Apple (d) Plum

2183. M-9 a dwarfing rootstock of apple has originated as a result of:

(a) Cross of M16 x M27

(b) Cross of M27 x Northern spy

(c) Cross of Northern spy x M16

(d) Chance seedling selection

2184. The ill effect of submergence of roots of deciduous fruit plant in water for long time is primarily due to :

(a) Lack of nutrition

(b) Lack of aeration

(c) Excess of moisture

(d) Poor absorption of moisture

2185. Domestic quarantine exist in India for :

(a) **Potato eelworm** (b) Sanjose scale

(c) Panama wilt (d) Sugarbeet eelworm

2186. Mulching of banana plant with black polythene' reduces the population of weed because it:

(a) Restricted the supply of oxygen

(b) **Obstruct the light and hence the weed starve**

(c) Result of excessive moisture in root zone of weed

(d) None of these

2187. Grape are pruned in only once in North India while twice in South India. This difference is due to the fact that grapes :

(a) **Shed leaves in winter in Northern India while they remain evergreen in Southern India**

(b) Less prone to disease in north India while they are more prone to disease in South India

(c) Have early bearing in North India while late in South India

(d) Bear lightly in Northern India while they bear heavily in South India

2188. Multi storeyed cropping practiced in :

(a) J & K and Kerala

(b) Bihar and Madhya Pradesh.

(c) **Karnataka and Kerala**

(d) Rajasthan and Haryana

2189. Litchi does not flower in South India in winter months because of:

(a) **Chilling requirement does not meet**

(b) Decrease of high temperature prevalent at the time of flower formation

(c) High RH at the time of flower formation

(d) The excess light stimulus present at the time of flower formation

2190. Which of the following fruits is cured in smoke for ripening?

(a) Bael (b) Karonda

(c) Banana (d) Papaya

2191. The nutrient element which improves quality of the fruit is :

(a) K (b) P

(c) S (d) N

2192. Guava wilt is caused due to :

(a) *Fusarium solani* (b) *Macrophomina phaseoli*

(c) *Cephalosporium* spp. **(d) *Gliocladium roseum***

2193. Pineapple is a native of :

(a) Africa (b) North America

(c) Brazil (d) Europe

2194. Papaya variety resistant to wind is :

(a) Pusa Delicious (b) Pusa Surya

(c) Pusa Giant **(d) Pusa Nanha**

2195. First President of Horticultural society of India was :

(a) Dr. G. S. Randhawa (b) Dr. R. N. Singh

(c) Dr. K. L. Chadha **(d) Dr. G. S. Cheema**

2196. Best soil for planting a fruit Orchard should have pH between:

(a) 6.5 - 7.5 (b) 7.5 - 8.0

(c) 6.0- 6.5 (d) 5.5 - 6.5

2197. *Zizyphus numularia* is known to impart dwarfness in ber, Dwarfness is due to :

(a) Graft incompatibility (b) Poor nutrient uptake

(c) Poor root formation (d) Virus infection

2198. Wind break for protecting orchards from strong winds is planted on which direction of the orchard:

(a) South-East (b) East-West

(c) South-West **(d) North-West**

2199. Wind break is effective upto the distance of (times) of its height?

(a) Two (b) Ten

(c) Eight **(d) Four**

2200. Edible part of Rambutan is :

(a) Placenta **(b) Aril**

(c) Mesocarp (d) Endocarp

2201. Present (2007) DDG (Deputy Director of General) of Horticulture-

(a) Dr. H. P. Singh (b) Dr. Sant Ram

(c) Dr. S. N. Pandey (d) Dr. G. Kalloo

2202. L-49 is a selection from Allahabad Safeda made by Dr. G. S. Cheema at -

(a) Delhi (1930) (b) Lucknow (1935)

(c) Allahabad (1929) **(d) Pune (1927)**

2203. Vegetatively propagated plants from a single mother plant which retain their characters after repeated multiplication are called as :

(a) Clone (b) Variety

(c) Strain (d) Cultivar

2204. Artificial hybridization in fruit crop in the world was first practiced by -

(a) Fairchild (b) Knight

(c) **Thomas Andrew Knight** (d) Camoreous

2205. Acta Horticulture is a publication of

(a) Japan Society of Horticultural Sciences

(b) International Society for Horticulture Science

(c) Horticultural Society of India

(d) American Society of Horticultural Sciences

2206. 'Sloh' is a hybrid between

(a) Peach x Apricot (b) Plum x Apricot

(c) Almond x Peach **(d) Peach x Almond**

2207. Inarching in mango in India was suggested by :

(a) Dr. R. N. Singh (b) Dr. P. K. Mukherjee

(c) Dr. K. L. Chadha **(d) Dr. S. K. Mukherjee**

2208. Outer layer of papaya seed which hinders germination is called?

(a) Sarcotesta (b) Festa

(c) Embryo (d) Integument

2209. Perlette, a early maturing grape variety was introduced in India from USA. This variety is bred by

(a) A.J.Winkler **(b) H. P. Olmo**

(c) S. S. Negi (d) A.B.Frost

2210. Based on mode of pollination tall Coconut cultivars are -

(a) Monogamous (b) Polygamous

(c) Allogamous (d) Autogamous

2211. Only citrus fruit which contain mallic acid :

(a) Pummelo **(b) Sweet lime**

(c) Sweet orange (d) Acid lime

2212. Type of banana which have persistent male bud at ripening :

(a) Horn plantain (b) Pseudo-horn plantain

(c) French plantain (d) None of these

2213. Five Corner fruit/Star fruit is known as :

(a) Phalsa (b) Persimmon

(c) Carambola (d) Rambuton

2214. Which form of sugar is predominantly present in grape at ripening?

(a) Glactose **(b) Glucose**

(c) Fructose (d) Sucrose

2215. High production in protected cultivation is due to :

(a) High temperature (b) High humidity

(c) High photosynthetic efficiency (d) All of the above

2216. Pistachionut is in nature.

(a) Monoecious (b) Polygamous

(c) Dioecious (d) None of these

2217. High profitable intercrop in Coconut and Arecanaut is :

(a) Pepper (b) Banana

(c) Tapioca (d) Pineapple

2218. 'Black Heart' is a variety of

(a) Strawberry **(b) Cherry**

(c) Blueberry (d) Raspberry

2219. Mangosteen *(Garcinia mangostana)* flower is sessile and botanically its fruit is :

(a) Berry (b) Drupe

(c) Nut (d) None of the above

2220. Mutation is a sudden change in the genetic makeup of any organism. It is

(a) Organal cell (b) Multicellular overy

(c) A single cell event (d) All of the above

2221. Kinnow a hybrid between of *Citrus nobilis* x *Citrus deliciosa* was developed by H. B. Frost in 1935 at :

(a) Australia **(b) California**

(c) Brazil (d) India

2222. A disease which can transform a "on year" of fruiting to "off year" in mango is

(a) Anthracnose (b) Leaf blight

(c) Stem cancer **(d) Powdery Mildew**

2223. Beneficial fungi associated with roots of litchi is -

(a) **Mycorrhizae** (b) Rhizobium

(c) Pythium (d) Phytophthora

2224. Mango variety Amrapali is regular bearer and suitable for close planting. What is demerit of this variety?

(a) Uneven fruit size (b) Mango malformation

(c) Both (a) and (b) (d) None of the above

2225. Major bottleneck in breeding of banana is :

(a) Pollen sterility (b) Non-receptivity of stigma

(c) Vegetative parthenocarpy (d) All of the above

2226. Early post blossom fruit drop in apple is known as :

(a) August drop (b) July drop

(c) June drop **(d) May drop**

2227. Nectarine is smooth skinned peaches. It has originated from Fuzzy skinned peaches as a result of mutation. The mutation is -

(a) Single dominant gene (b) Recessive

(c) Multiple allels (d) None of these

2228. Saharanpur Seedless is a seedless variety of guava. The seedlessness is due to :

(a) Vegetative **(b) Triploidy**

(c) Stimulative (d) None of these

2229. Ficin (Rennin) has 30-100 times more milk clotting activity than animal rennin. This is present in latex of which fruit crop :

(a) Fig (b) Apple

(c) Pear (d) Peach

2230. 'Mangosteen' fruit contains seed. The development of mangosteen seed is from nucellar tissue hence propagates true to type through sexual propagation. This process of seed formation is called:

(a) Parthenocarpy (b) Polyembryony

(c) Parthenogenesis (d) Apomixis

2231. Pineapple is a multiple fruit, formed of number of fruitlets as a result of fusion of inflorescence. Each fruitlet of pineapple is a:

(a) Drape (b) Berry

(c) Achene (d) None of these

2232. A series of biochemical changes, initiated by the catalytic production of ethylene, making the changes from growth to senescence and involving an increase in respiration rate is called:

(a) Maturity (b) Senescence

(c) Climacteric (d) Ripening

2233. Anab-e-Shahi a seeded grape cultivar was selected in 1930 from the collection of Nawab Baquer Ali Khan, who was associated with its selection :

(a) Nawab Baquer Ali Khan **(b) R. S. Pillay**

(c) Arve (d) None of these

2234. Pulp colour in guava is controlled by monogenically. What is correct about the inheritance of red pulp colour in guava:

(a) Red and white colour is co-dominant

(b) Red pulp colour is recessive

(c) White colour is dominant

(d) Red pulp colour is dominant over white pulp guava

2235. 'Oroblanco' is a triploid (3x) citrus originated as result of cross between:

(a) Ruby Orange x King Tangore

(b) Trifoleate Orange x Washington Naval Orange

(c) Trifoliate Orange x Ruby Orange

(d) None of these

2236. Sex in grape is under the control of two alleles. Femaleness is governed by (Sf) and hermaphorodite by (Sh). What is the inheritance pattern of sex in Grape?

(a) Sf is dominant **(b) Sh is dominant**

(c) Sh is recessive (d) Sf is recessive

2237. Persimmon was introduced in 1921 in KulJu vally of Himachal Pradesh from Japan. Who was associated with its introduction in India?

(a) Alexander Coult (b) S. N. Stokes

(c) P. K. Mathur **(d) A. N. Lee**

2238. Cell division continues till maturity in fruit development of:

(a) Peach (b) **Avocado**

(c) Strawberry (d) Mango

2239. Shoot tip grafting is used to produce :

(a) Virus free plant (b) Salt tolerant plant

(c) Colourful plant (d) Dwarf plants

2240. Which hormone is more effective in hulling of walnut?

(a) GA_3 **(b) Kinetin**

(c) IAA (d) Ethephon

2241. Most acidic citrus fruit belongs to :

(a) *Citrus meduransis* **(b) *Citrus megalaxycarpa***

(c) C. *Jambhiri* (d) C. *Pennivesculata*

2242. Homozygosity in early generation is fast due to :

(a) Sub-mating (b) Hybridization

(c) Selfing (d) Back crossing

2243. The oldest fruit known to mankind is :

(a) Rambuton (b) Persimmon

(c) Date palm (d) Sapota

2244. Which one of the following fruit crops has maximum diversity?

(a) Citrus (b) Banana

(c) Apple (d) None of these

2245. Coppery pigmentation in Guava is due to deficiency of :

(a) K20 (b) N

(c) P_2O_5 (d) None of these

2246. In which citrus fruit, vivipary is found ?

(a) C. *grandis* **(b) C. *paradisii***

(c) C. *limon* (d) None of these

2247. Which mango variety retain flavour after processing?

(a) Amrapali (b) Mallika

(c) Ambika **(d) Alphanso**

2248. Why Paclobutrazol used as a anti GA :

(a) It inhibit only growth of the plant no colour and flavour

(b) It inhibit growth, colour and flavour

(c) It doesn't effect on any growth, colour and flavour

(d) None of these

2249. Which fruit crop have high (more) alternate bearing?

(a) Mango (b) Avocado

(c) Olive (d) None of these

2250. Phalsa is a :

(a) Single seeded berry **(b) Multi-seeded berry**

(c) Two seeded berry (d) None of these

2251. Viral disease transmitted by seed is :

(a) Psorosis (b) Canker

(c) Exanthema (d) Die-back

2252. Among the Indian States, highest producer of Banana is:

(a) Maharashtra **(b) Tamil Nadu**

(c) West Bengal (d) Gujarat

2253. In which of the following fruit, Thailand is the leading producer in the world?

(a) Mango (b) Banana

(c) Grape **(d) Pineapple**

2254. Highest productivity of grape is in :

(a) Italy **(b) India**

(c) USA (d) China

2255. Which of the following variety of mango is monoembryonic?

(a) Olour (b) Goa

(c) Rataul (d) Salem

2256. Which of the following fruits is deciduous in nature?

(a) Bael (b) Loquat

(c) Ber (d) Litchi

2257. Identify the fruit belonging to non-climacteric group :

(a) Persimmon (b) Avocado

(c) Pineapple (d) Mango

2258. In which group you classify pineapple, according to its respiration rate?

(a) Low (b) Moderate

(c) High (d) Very high

2259. In which group you classify mango, according to its ethylene production rate?

(a) Very low (b) Low

(c) Moderate (d) High

2260. Jackfruit is :

(a) Anemophilous (b) Entomophilous

(c) Parthenophilous (d) Self-pollinated

2261. The breeding methodology used for developing seedless grape is:

(a) Double haploidy (b) Aneuploidy

(c) **Triploidy** (d) Autopolyploidy

2262. Non-climacteric fruits are those which :

(a) Will ripen properly only while attached to the plant

(b) Will ripen properly only when harvested from the plant before ripening

(c) Will slow an abrupt increase in respiration rate at the ripening stage

(d) Do not have big seeds

2263. Sigmoid growth curve in pineapple is :

(a) Single (b) Double

(c) Triple (d) None

2264. Ambika is a Hybrid variety of Mango crossed within :

(a) Amrapali x CISH M5

(b) Amrapali x Janardan Pasand

(C) Neelum x CISH MI

(D) Neelum x Janardan Pasand

2265. "Pusa Srijan" a dwarfing root stock of Guava has been released by:

(a) CISH (b) IIHR

(c) CITH **(d) IARI**

2266. " Pusa Arunima " is a hybrid variety of Mango recently released from IARI is a cross between :

(a) Amrapali x Sensation

(b) Amrapali x Janardan Pasand

(c) Sensation x Amrapali

(d) Sensation x Janardan Pasand

2267. "Pusa Surya" is a variety of Mango recently released from IARI, is a :

(a) Hybrid **(b) Selection**

(c) Mutant (d) None of these

2268. Genetic Constitution of Pusa Srijan is :

(a) 2n + 2 (b) 2n – 2

(c) 2n + 1 + 1 **(d) 2n - 1 - 1**

2269. India's share in global production of fruit is :

(a) 10% (b) 20%

(c) 5% **(d) 15%**

2270. Pollination is not required for fruit set in :

(a) Mango (b) Ber

(c) Banana (d) Cherry

2271. Which particular variety of *Sweet Orange* has its minimum acid content?

(a) Jaffa **(b) Mosambi**

(c) Blood Red (d) Hamlin

2272. The edible portion of Cashew apple is modified of :

(a) Ovary (b) Thalmus

(c) Peduncle (d) Calyx

2273. What is propagation method of Mangosteen?

(a) Patch budding (b) Tongue Grafting

(b) Seed (d) Cuttings

2274. Haploid chromosome number of Litchi is :

(a) 11 (b) 12

(c) 10 **(d) 15**

2275. L-49 is a variety of Guava developed at:

(a) Lucknow (b) Allahabad

(c) Pune (d) Delhi

2276. 'Temple' is a rootstock of :

(a) Grape (b) Ber

(c) Phalsa (d) Citrus

2277. Cage technique is used for :

(a) Papaya breeding (b) Grape breeding

(c) Mango breeding (d) None of these

2278. *'Chok Throat'* in banana is due to :

(a) Low temperature (b) High humidity

(c) Low humidity (d) None of these

2279. Ascorbic Acid content in Guava is maximum in :

(a) Pulp **(b) Peel**

(c) Seed (d) Core

2280. Which of the following grape variety is *tenturier* in nature?

(a) Arkavati (b) Pusa Urvashi

(c) Arka Sweta **(d) Pusa Navrang**

2281. Bottom Heat Treatment of mango cuttings from roofings is done at:

(a) 50°C (b) 32°C

(c) 70 °C (d) 100 °C

2282. *Sonaka* is a clonal selection of :

(a) Thompson Seedless (b) Pusa Seedless

(c) Beauty Seedless (d) Kishmis Charni

2283. The most important hindrance in fruit breeding is :

(a) Long juvenile phase (b) Less seed per fruit

(c) Large size tree (d) Vegetative propagation

2284. Which of the following chemical is used for polyploidy breeding?

(a) MH (b) NAA

(c) Cholchicine (d) SADH

2285. Which of the following technique is used for detection of virus?

(a) APD (b) RFLP

(c) AFLP **(d) ELISA**

2286. Critical operation in Banana production is :

(a) Proping (b) Desuckering

(c) Weeding (d) Mulching

2287. Soil application of *"cultar"* is recommended in mango orchard. Its commercial dose is :

(a) 4-5 g/tree (b) 0.4- 0.5 g/tree

(c) 40 - 50 g/tree (d) None of these

2288. NRC on Grape is located at:

(a) Rahuri (b) **Pune**

(c) Nagpur (d) Ludhiana

2289. Central Institute of Sub-tropical Horticulture (CISH) is located at:

(a) UP (b) Delhi

(c) Maharashtra (d) AP

2290. Normally on which side of mango tree, panicles emerge first

(a) Eastern and S. Eastern (b) Northern

(c) North-East (d) Western

2291. Indian Journal of Horticulture is a publication of :

(a) ICAR (b) IARI

(c) HSI (d) IIHR

2292. Leading foreign earning crop in India is :

(a) Mango (b) Apple

(c) Banana **(d) Cashew**

2293. Soil Erosion can be efficiently checked by :

(a) Growing plants with extensive fibrous root system

(b) Planting tall growing trees at wider spacing

(c) Dividing the farm land into several small plots

(d) Farming the land with a gentle slope

2294. High water table is a limiting factor in the cultivation of :

(a) Pear **(b) Peach**

(c) Plum (d) None of these

2295. Which of the following trees require pruning regularly?

(a) Sapota **(b) Peach**

(c) Strawberry (d) All of above

2296. Recombination of genes takes place:

(a) **During meiosis I** (b) During meiosis II

(c) On completion of fertilization (d) During all above stages

2297. The most important method of propagation in apple is :

(a) T budding (b) Air layering

(c) Inarching (d) **Bench grafting**

2298. A plant species has 6 chromosomes in each of the 4 cells of its pollen treated. The number of chromosomes present in its leaf tips would be:

(a) 3 (b) 6

(c) **12** (d) 24

2299. Grafting can be successful if there is fusion of :

(a) Phloem (b) **Cambium**

(c) Xylem (d) Campanion cells

2300. The hormone responsible for apical dominance is :

(a) Abscissic acid (b) **Auxin**

(c) Zeatin (d) Ethylene

2301. The practice of cutting away a portion of plant to strike a balance between vegetative growth and production is known as :

(a) Notching (b) Girdling

(c) Training (d) **Pruning**

2302. In order to avoid low temperature injury the temperate peaches should be pruned during :

(a) **Mid of February** (b) First week of January

(c) Last week of January (d) Late December

2303. How many maximum seeds are possible in an apple fruit ?

(a) 5 (b) **10**

(c) 12 (d) 15

2304. Which portion of the fruit is consumed in pome fruits ?

(a) Thalmus (b) Outer growth of seed
(c) Aril (d) Inflorescence

2305. Pistachionut (P. *vera)* belongs to family :

(a) Rubiaceae (b) Juglandaceae
(c) Anacardiaceae (d) None of these

2306. Walnut is commercially propagated through :

(a) Chip budding **(b) Patch budding**
(c) Both (a) and (b) (d) None of these

2307. Apple shows which type of Apomixis ?

(a) Non recurrent **(b) Recurrent**
(c) Budding (d) Vegetative

2308. Inarching is practiced to repair the roots whereas to repair stem which type of grafting is used?

(a) Bridge grafting (b) Approach grafting
(c) Bark grafting (d) Bench grafting

2309. Basic chromosome number of pomoideae is :

(a) 7 (b) 19
(c) 17 (d) 18

2310. "Top Red" is a bud sport of which variety ?

(a) Golden Delicious **(b) Starking Delicious**
(c) Red Gold (d) Red Delicious

2311. Aroma in the ripe fruit of Apple is due to :

(a) Hexanol (b) Eugenol
(c) Isopentanol **(d) Ethyl-2-Methyl butyrate**

2312. "Mahal" a rootstock of Pear is scientifically known as :

(a) *Pyris pashia* (b) *P. pyrifolia*
(c) *P. serotina* (d) *P. communis*

2313. Self unfruitful variety of Peach is :

(a) **J. H. Hale** (b) July Elberta

(c) Sharbati (d) Shan-e-Punjab

2314. "Water Core", a disorder of apple is formed due to :

(a) Water stress

(b) Water excess

(c) High temperature

(d) Formaton of Ice crystal and deposition of sorbitol

2315. Almond principally bears on the :

(a) Old **(b) Spurs**

(c) New shoot shoots (d) Current season growth

2316. Which variety of Banan is used for making baby food ?

(a) Nendran (b) G-9

(c) Poovan (d) Hill Banana

2317. The main objective of growing intercrop with the main fruit crop is :

(a) Improve soil fertility

(b) **Check soil erosion**

(c) Help the fruit crop to grow better

(d) None of these

2318. Micro nutrient loving plant is :

(a) Apple (b) Banana

(c) Citrus (d) Papaya

2319. M-27 a dwarfing rootstock of apple is cross between :

(a) M9 x M13 (b) M12 x M16

(c) M12 x M15 (d) M20 x M9

2320. Walnut a monoecious fruit and its pollination is through :

(a) Animals (b) Insects

(c) Birds **(d) Wind**

2321. Which type of dichogamy occurs in walnut?
(a) Protogyny
(b) Protandry
(c) Heterostyly
(d) None of these

2322. Which crop has energy value twice as much as banana fruit?
(a) Apricot
(b) Avocado
(c) Apple
(d) Almond

2323. "Colt" is a rootstock of :
(a) Pear
(b) Plum
(c) Apple
(d) Cherry

2324. "False bloom", is a disorder associated with excessive water. It is observed in which of the following fruit crops?
(a) Strawberry
(b) Craneberry
(c) Raspberry
(d) None of these

2325. Low-chilling peach variety is :
(a) Kanto-5
(b) Shimizo Hakuto
(c) Shan-e-Punjab
(d) July Elberta

2326. Low-chilling plum variety is :
(a) Santa Rosa
(b) Titron
(c) Beauty
(d) Methley

2327. Fruit crop resistant to wet soil is :
(a) Peach
(b) Apricot
(c) Quines
(d) Pear

2328. A dwarf rootstock of pear is :
(a) Quince A
(b) Quince B
(c) Quince C
(d) Kainth

2329. Cell division continues till maturity in fruit development of :
(a) Peach
(b) Strawberry
(c) Avocado
(d) Apple

2330. In papaya, cool climate results into :

(a) **Fruits attain small size** (b) Flowers and fruit drop

(c) Reduction in T.S.S. (d) Green colour appearance

2331. Bitterness in peach is due to :

(a) **Prunasin acid** (b) Sugar

(c) Maleic acid (d) Hydrocynadin

2332. First successful transgenic fruit plants are produced in :

(a) **Pear** (b) Cherry

(c) Walnut (d) Apple

2333. Sigmoid growth curve in kiwi fruit is :

(a) Single sigmoid (b) Double sigmoid

(c) **Triple sigmoid** (d) No growth curve is found

2334. Walnut is a native of :

(a) India (b) **Iran**

(c) Afghanistan (d) Pakistan

2335. Nemaguard', a nematode is resistant roostock of :

(a) Plum (b) **Pear**

(c) Cherry (d) Apple

2336. Standard unit used in expressing refrigeration capacity is :

(a) Calory (b) Amperes

(c) **Ton** (d) Jule

2337. Ideal storage temperature of banana is :

(a) 7 - 10 °C (b) **13 - 14 °C**

(c) 1-3 °C (d) 5-6 °C

2338. Litchi takes months to mature?

(a) 3.5-4.5 months (b) **2.5 to 3.5 months**

(c) 4-5 months (d) None of these

2339. The pigment present in Jamun is :

(a) Carotenoid **(b) Anthocyanin**

(c) Lycopens (d) Xanthophyll

2340. Kharna Khatta is a commercial rootstock of citrus preferred in light soil of:

(a) Punjab (b) MP

(c) Karnataka (d) Uttar Pradesh

2341. Which of the following adversely affects the storage?

(a) Ethylene (b) Auxin

(c) Gibberellin (d) Kinetin

2342. Filbert can be propagated through :

(a) Tip layering (b) Compound layering

(c) Stool layering (d) Simple layering

2343. Grape has originated in the area extending from Black Sea and Caspian Sea but was introduced in India from :

(a) Australia **(b) Afghanistan**

(c) Brazil (d) U.S.A.

2344. Acid lime is propagated by :

(a) Grafting **(b) Budding**

(c) Cutting (d) Seed

2345. Mango pulp is preserved by :

(a) KMS (b) Citric acid

(c) Sodium Benzoate (d) None of these

2346. Guava is utilized maximum for the manufacture of :

(a) Jam (b) Marmalade

(c) Jelly (d) Juice

2347. If irrigation is not available which aspect (elevation) is suitable for apple cultivation ?

(a) Northern (b) Southern

(c) Both (a) and (b) (d) None of these

2348. In apple, which rootstock have maximum root suckers ?

(a) M-7 (b) MM-106

(c) M-5 (d) MM-103

2349. In apple contour planting system is followed if slope is :

(a) less than 10% (b) 10%

(c) More than 10% (d) None of these

2350. Generally peach trees are intercropped in apple plantation because:

(a) Short juvenile period and tree size is more

(b) Long juvenile period and tree size is more

(c) **Short juvenile period and tree size is smaller**

(d) None of these

2351. Strawberry grow best at:

(a) High hills **(b) Mid hills**

(c) Low hills (d) Very high hills

2352. Banana seeds are in nature.

(a) Soft (b) hard

(c) **very hard** (d) None of these

2353. Pollen viability is determined by :

(a) Bromine test (b) **Aceto-carmine test**

(c) Salt-acid test (d) None of these

2354. Flower Bud Differentiation (FBD) of Mango in florida :

(a) October (b) December

(c) March (d) August

2355. Special character of "False Horn" is :

(a) To produce three bunches (b) **To produce two bunches**

(c) not to produce any bunch (d) None of these

2356. Totapari Red Small (TRS) has been found to have very short juvenile phase and thus it can be used as a male parent in hybridization programme to reduce the length of juvenile phase : Why not used as a female parent ?

(a) Female parent cannot be used in hybridization

(b) When used as a famale parent it exhibit long juvenile phase

(c) **Because there is no effect of the female parent has been found on the distribution of the juvenile period**

(d) None of these

2357. Sindhu (a variety of mango) is evolved from :

(a) Vengurla, Maharashtra (b) Pune, Maharashtra

(c) Nagpur, Maharashtra (d) None of these

2358. Which variety of mango has very less storage period?

(a) Amrapali (b) Mulgoa

(c) Totapari (d) Neelum

2359. Skiffing practice is used in :

(a) Rubber (b) Cocoa

(c) Coffee (d) Tea

2360. Temperature suitable for coconut growth is :

(a) 21 °C (b) 29°C

(c) 27°C (d) 24 °C

2361. Precursor of ethylene is :

(a) Tryptophane **(b) Methionene**

(c) Purine (d) None of these

2362. 'Mridula' is a variety of :

(a) Pomegranate (b) Phalsa

(c) Cashew (d) Karonda

2363. Origin place of pomegranate is ;

(a) Iran (b) Iraq

(c) Afghanistan (d) India

2364. An example of slow release fertilizer is :
(a) CAN (b) KCl
(c) DAP **(d) Lac Coated Urea**

2365. After inflourescence emergence which fruit take more time for mature :
(a) Aonla (b) Custard Apple
(c) Banana (d) Pomegranate

2366. "Black Pod" is a disease of :
(a) Arecanut (b) Coconut
(c) Cashewnut **(d) Cocoa**

2367. In North India, which mandarin is mostly grown ?
(a) Lemon **(b) Kinnow**
(c) Acid lime (d) Pummelo

2368. Fruit thinning growth regulator:
(a) IA A (b) CK
(c) NAA **(d) GA**

2369. Father of "Golden Revolution" is :
(a) Dr. K. L. Chadha (b) Dr. M. S. Swaminathan
(c) Dr. V. Kurian (d) Dr. M. Mahadevappa

2370. NRC for Litchi is situated at:
(a) Ludhiana, Punjab (b) **Muzaffarpur, Bihar**
(c) Pune, Maharashtra (d) Saharanpur, Uttar Pradesh

2371. NRC for Pomegranate is situated at:
(a) Parbhani, Maharashtra (b) Nagpur, Maharashtra
(c) Solapur, Maharashtra (d) Pune, Maharashtra

2372. The seeds of a transgenic plant will not be to germinate in next time?
(a) Due to pleiotropic effect (b) Gene silencing
(c) Methylation (d) Delayed segregation

2373. The variety of mango have high TSS is :

(a) Amrapali (b) Dashehari

(c) **Sahib Pasand** (d) Gulab Khas

2374. In which fruit crop 'Disapping' practice is used,

(a) Annona **(b) Mango**

(c) Papaya (d) Sapota

2375. The antisterility Vitamin is :

(a) Vitamin K (b) Vitamin D

(c) Vitamin A **(d) Vitamin E**

2376. To regulate the cropping the best time of pruning of guava shoots in UP plains is :

(a) April-June (b) December-January

(c) October - November **(d) February - March**

2377. Kinnow Mandarin was developed by :

(a) G. S. Cheema (b) R. N. Singh

(c) T. K. Bose **(d) H. B. Frost**

2378. Which insect leaves stylet sheath during feeding :

(a) Brown Plant Hopper (b) Gall Midge

(c) Case Worm (d) Yellow Stem Borer

2379. Iron deficiency in plant is characterized by :

(a) Little leaf

(b) **Intervenial chlorosis of old leaves**

(c) Yellowing of new leaves

(d) Mottling of leaf

2380. Ideal time for pruning of phalsa is :

(a) December - January (b) January - February

(c) February March (d) November-December

2381. Which of the following disease is transmitted by vector?
(a) Greening (b) Exocortis
(c) Little leaf **(d) Teristeza**

2382. Best rootstock for citrus in South India is :
(a) Cleopatra mandarin (b) Citrange
(c) Rangpur lime (d) None of these

2383. Tangelo is a cross between :
(a) Pummelo x Tangerin **(b) Tangerin x Pummelo**
(c) Pummelo x Grape fruit (d) Grape fruit x Sweet orange

2384. The term ' Pomology' has been derived from :
(a) Greek (b) Latin
(c) German (d) Italy

2385. Bitterness in citrus juice is due to :
(a) Acid **(b) Limonin**
(c) Glucosides (d) Acid : Sugar Ratio

2386. High cytokinin/auxin ratio in callus favours:
(a) Root development (b) Flowering
(c) Somatic embryogenesis **(d) Shoot formation**

2387. On tree storage in citrus is possible with application of :
(a) 2,4-D (b) Thiourea
(c) ABA **(d) GA3**

2388. Which of the following is tetraploid fruit crop?
(a) Ber (b) Citrus
(c) Papaya (d) Plum

2389. Little leaf in citrus pertains to deficiency of :
(a) Chloride (b) Sulphur
(c) Copper **(d) Zinc**

2390. The suitable pair of fruit crop for a soil of 9.5 pH is :

(a) Aonla and Jamun (b) Aonla and Guava

(c) Jamun and Date palm (d) Date palm and Aonla

2391. Grape is essentially a :

(a) Temperate fruit plant (b) Tropical fruit plant

(c) Sub-tropical fruit plant (d) All of the above

2392. The fruit which have highest lenticle is :

(a) Apple (b) Apricot

(c) Pear (d) Plum

2393. Most susceptible fruit crop for water logging is :

(a) Banana **(b) Papaya**

(c) Guava (d) Aonla

2394. The varieties of mango resistant to powdery mildew are :

(a) Latif, Hur and Sinduri

(b) Latif, Amrapali and Totapuri

(c) Totapuri, Sinduri and Amrapali

(d) Hur, Sinduri and Totapuri

2395. Dwarfism in banana plant is governed by :

(a) Single dominant gene (b) Recessive genes

(c) Multiple genes (d) None of these

2396. In grape which training system gives maximum cost benefit ratio?

(a) Kniffin **(b) Bower**

(c) Telephone (d) Head

2397. Which disease mostly occurs in high density planting of banana:

(a) Panama wilt **(b) Finger tip**

(c) Bunchy top (d) Sigatoka leaf spot

2398. As per latest record number of species in mango are :

(a) 58 (b) 72

(c) 69 (d) 52

2399. Barbados cherry, the richest source of Vitamin C belongs to family:

(a) Annonaceae **(c) Malphigiaceae**

(b) Apocynaceae (d) None of these

2400. Citrus species having aroma similar to eucalyptus is :

(a) *Citrus Sinensis* (c) *Citrus reticulata*

(b) *Citrus arvensis* ***(d) Citrus assmensis***

2401. Date palm plant bear

(a) Unisexual (b) Bisexual

(c) Monodioecious flowers (d) None of these

2402. Grape like aroma in Jamun is due to the presence of :

(a) Easter compound (b) Anthocyanin

(c) Methyl anthranilate (d) Tarteric acid

2403. "Head in fire" and "foot in water" is associated with :

(a) Tamrind (b) Aonla

(c) Cashew **(d) Date palm**

2404. Which of the typical temperate fruit

(a) Pecan nut **(b) Apple**

(c) Cherries (d) Walnut

2405. Which of the male sterile variety of peach

(a) Elberta (b) Florda queen

(c) J.H.Hale (d) Maygold

2406. Which of the one following variety of nectarine?

(a) Veteran (b) Condor

(c) Silver lode (d) Rio Oso Gem

2407. Maturity stage of apple?

(a) L stage **(b) T stage**

(c) P stage (d) Z Stage

2408. Best time of pruning in peach.

(a) January (b) March

(c) December (d) October

2409. Dwarfing rootstock of pear.

(a) Quince A **(b) Quince C**

(c) Quince B (d) Kainth

2410. Blueberries belong to the family?

(a) Fagaceae **(b) Ericaceae**

(c) Ebenaceae (d) Rosaceae

2411. Which of the following is not Indian variety?

(a) Glister (b) Ambri

(c) Ambrich (d) Ambryol

2412. Waxing on fruit is done to reduce ?

(a) Transpiration (b) Respiration

(c) Evaporation (d) Reducing ripening

2413. Colour in apple is developed?

(a) UV-rays (b) Visible rays

(c) Infra-red (d) All

2414. Colt and Mazzard & F12 is a rootstock?

(a) Pear (b) Apple

(c) Cherry (d) peach

2415. Heavily pruned crop from the following?

(a) Apple **(b) Peach**

(c) Plum (d) pear

2416. Semi-vigourous rootstock in cherry

(a) Paja **(b) Mazard F-12/1**

(c) Colt (d) St. Julian

2417. Virus resistant rootstock of apple?

(a) EMLA (b) Malling matan

(c) Ottowa (d) All

2418. Most winter hardy rootsock in apple?

(a) M9 (b) MM126

(c) MM104 (d) M26

2419. M-26 cross is?

(a) M6×25 (b) Northen Spy×M9

(c) M9×M16 (d) Northern Spy×M2

2420. Which part of the decline of pear plant?

(a) Xylem (b) Shoot

(c) Phloem (d) Trunk

2421. Peach Variety Florida Queen developed by?

(a) R.H. Sharpe (b) Dr K.K. Jindal

(c) R.N. Singh (d) D.P. Singh

2422. Which fruit most sorbital is found?

(a) Peach **(b) Apple**

(c) Almond (d) Walnut

2423. Which of the one following Apple new rootstock?

(a) Pusa Seb mulbrint 1(PSM) (b) MM 120

(c) Pusa mulbrint 2 (d) M1

2424. Which of the variety of walnut dicogamy is absent?

(a) Vina **(b) Payne**

(c) Serr (d) Hertly

2425. Botanical name of pecannut?

(a) *M. serrata* (b) *V. myrtittoides*

(c) *Carya illinoenssis* (d) All

2426. Leading commercial variety of apple in export and import?

(a) Fuji and gola (b) All

(c) Ambri (d) Organ spur

2427. Best time grafting of apple?

(a) March-April (b) Dece.nber -January

(c) January-Feb (d) Feb-March

2428. Type of plum require maximum pruning.

(a) Japanese (b) European

(c) Amarican (d) All

2429. Multirow & bed row developed for?

(a) Apple (b) Plum

(c) Peach (d) Pear

2430. For temperate fruit scion buds are stored at?

(a) 35^0f **(b) 40^0f**

(c) 32^0f (d) 45^0f

2431. Apple var Sunhari is crossed b/w?

(a) Ambri×Golen delicious (b) Royal delicious×Ambri

(c) Ambri × Red delicious (d) Red delicious×Ambri

2432. Apomixis in apple

(a) Obligated (b) facultative

(c) Non recurrent (d) None

2433. Redness in apple due to?

(a) Lycopene **(b) Anthocyanin**

(c) Carotene (d) xanthophyll

2434. Aroma in the ripe of fruit of apple due to?

(a) **Ethyl-2 methyl** (b) Eugenol

(c) Isopentanol (d) Hexanol

2435. Why famous Ambri var?

(a) All (b) **Long stroge life**

(c) Expot quality (d) Good quality

2436. Which fruit crop resistant to Mn deficiency?

(a) **Strawberry** (b) Almond

(c) Peach (d) Walnut

2437. King of nut?

(a) Pecannut (b) **Walnut**

(c) Arecanut (d) chest nut

2438. Apricots is originated?

(a) western china (b) **E. china**

(c) None (d) S. Amarica

2439. Low chilling apple var?

(a) **Tropical beauty** (b) Mutsu

(c) Fuji (d) Gola

2440. Which of the one following European type of pear?

(a) Max red Bartlett (b) Flemish beauty

(c) **Bartlett** (d) All

2441. Which of the one following Asian type of peach?

(a) Kiffer (b) Pathar-nakh

(c) Le-conte (d) **All**

2442. Which of the one following low chilling apple var.?

(a) Michal (b) Neomi

(c) Tamma (d) **All**

2443. Nugget is var of?

(a) Walnut **(b) Apricot**

(c) Almond (d) Apple

2444. Which of the one following Japanese plum?

(a) Santa rosa (b) President

(c) Starking delicious (d) Green gage

2445. Which of the one following European plum

(a) Santa rosa **(b) President**

(c) Beauty (d) Mriposa

2446. Which of the self fruitful var of Almonds

(a) Drake (b) Katha

(c) Dhebar **(d) All**

2447. Nectarine is fuzzless.

(a) Plum **(b) Peach**

(c) Pear (d) Apple

2448. India shares how many percentage of world mango cultivation?

(a) 80 (b) 50

(c) 60 **(d) 54**

2449. Which of the unfruitful cv. of peach

(a) J.H.Hale (b) Saharanpur prabhat

(c) Shaline (d) Sarbati

2450. Botanically apple is :

(a) Berry **(b) False fruit**

(c) Syconus (d) True fruit

2451. Which var is grit cell in pear?

(a) Flemish Beauty (b) Punjab gold

(c) Gola (d) Kiffer

2452. Queen of nut?

(a) Arecanut (b) **Pecannut**

(c) None (d) Walnut

2453. Which temperate fruit is earliest flowering?

(a) Almond (b) Apple

(c) Cherry (d) **Peach**

2454. Triple sigmoid curve?

(a) Apple (b) **Kiwifruit**

(c) peach (d) Walnut

2455. Which planting system is generally adopted in strawberry?

(a) **Matted row** (b) Double hill row

(c) Hill row (d) Undulated ridges

2456. Which of the not character of the apple?

(a) Deciduous (b) Alternet bearing

(c) **Polyemryony** (d) None

2457. Which of the highest chilling requirement of fruit ?

(a) **cherry** (b) Peach

(c) Apple (d) Plum

2458. Chilling requirement of bartlet pear ?

(a) 1000 (b) 1200

(c) **1500** (d) 750

2459. Stenospermocarpy is found?

(a) Plum (b) Apple

(c) All (d) **Peach**

2460. PPVFRA act in?

(a) 2000 (b) **2001**

(c) 2005 (d) 2003

2461. Which of the colour var. of apple?

(a) Topred (b) Red chif

(c) Vance delicious (d) All

2462. Oil content of seed found in peach?

(a) 40-50% (b) 50-60%

(c) 70-80% (d) 30-40%

2463. Leading producer of kiwi?

(a) Canada (b) Amarica

(c) None **(d) Italy**

2464. Which of the most late mature kiwi var of ?

(a) Abott **(b) Hayward**

(c) Allison (d) Bruno

2465. Which month mature time of apricot fruit?

(a) June to july **(b) July to august**

(c) None (d) May to june

2466. Hetrodicogamy is observed in ?

(a) Walnut (b) Almond

(c) Pecannut (d) All

2467. Pecannut belong to family?

(a) Juglandaceae (b) Actinidiaceae

(c) None (d) Rosaceae

2468. Strawberry is propagated through:

(a) Runner (b) Suckers

(c) None (d) Cutting

2469. June drop in apple is due to:

(a) Alternate bearing (b) Lack of pollination

(c) Moisture stress (d) Ca deficiency

2470. Apple seed stratification required temperature.

(a) **4^0C** (b) 5^0C
(c) 6^0C (d) 7^0C

2471. Bur knott formation is commonly found in.

(a) Pear (b) Plum
(c) Apple (d) Cherry

2472. Which of the one following apple scab resistant var of.

(a) Prima **(b) All**
(c) Red free (d) Freedom

2473. Black heart var of

(a) Apple **(b) Cherry**
(c) Peconnut (d) Walnut

2474. Apple was first introducing of India?

(a) 1887 (b) 1667
(c) 1927 (d) 1787

2475. Diploid cultivar of apple is?

(a) Self unfruitful (b) Cross fertile
(c) Self sterile **(d) Self fruitful**

2476. Pear is originated

(a) None **(b) Europe and Asia**
(c) China (d) Japan

2477. Prunes are dried?

(a) Peach (b) cherry
(c) Apricot **(d) Plum**

2478. Which type of pollination found in walnut.

(a) Animals **(b) Winds**
(c) Birds (d) Insect

2479. Hazelnuts belong to family?

(a) Ebenaceae (b) Proteaceae
(c) Betulaceae (d) Juglandaceae

2480. Which of triploid variety of apple?

(a) fuji **(b) Baldwine**
(c) Red chif (d) All

2481. Which of the edible portion of strawberry.

(a) Endocarp
(b) Bract & Perianth
(c) Mesocarp
(d) Receptacle with numerous achenes

2482. Which of the fruitful variety of cherry?

(a) Rubinaa **(b) Stella**
(c) Pazza (d) Colt

2483. Roopa is improved selection of?

(a) Walnut (b) None
(c) Almond (d) Peconnut

2484. Leaf curl is a serious disease of?

(a) Plum **(b) Peach**
(c) Walnut (d) Apple

2485. Strawberry flower is occurring?

(a) Single (b) Double
(c) Cluster (d) None

2486. Early var. of plum?

(a) Grand duke (b) Lombard
(c) Gullam **(d) Methley**

2487. Time of flowering in pear?

(a) January-Feb (b) March-April

(c) May- June **(d) Feb-March**

2488. Extremely late variety of pear?

(a) Weldon (b) Victoria

(c) Conference (d) Shinsui

2489. Chilling requirement of pear?

(a) 1200hrs (b) 700hrs

(c) 1500hrs (d) 500hrs

2490. Which is the polinizer var. of apple?

(a) Red delicious (b) Golden delicious

(c) All (d) Royal delicious

2491. Chromosome no of strawberry?

(a) 2n=56 (b) 2n=64

(c) 2n=48 (d) 2n=16

2492. Time of flowering in apple ?

(a) January-Feb **(b) Feb-March**

(c) None (d) March-April

2493. Planting distance of peach?

(a) 7×7 (b) 6×6

(c) 9×9 **(d) 5×5**

2494. Planting distance in strawberry?

(a) 35×35 **(b) 45×45cm**

(c) 40×40 (d) 30×30cm

2495. Yellow leaf disease found in?

(a) Arecanut (b) Walnut

(c) Almond (d) Peconnut

2496. Winter banana is variety of.

(a) Banana (b) None
(c) Apple (d) Strawberry

2497. Flower colour of peach.

(a) Pink (b) Blue
(c) White (d) Yellow

2498. Best time of pruning in peach?

(a) Spring **(b) Mid summer**
(c) Mid winter (d) None of above

2499. Which of the short day plant?

(a) Apple **(b) Strawberry**
(c) Peach (d) None of above

2500. Richest source of fat?

(a) Walnut (b) Almond
(c) Pecannut (d) None

2501. Which fruit suitable for cold desert of cold arid zone?

(a) Apple **(b) Apricot**
(c) Walnut (d) Cherry

2502. Apple scab is caused by?

(a) *Venturia inequalis* (b) *Dematophora necatrix*
(c) *Podoslhere leuchotricha* (d) None of these

2503. Type of incompatibility found in almond?

(a) Sporophytic (b) None
(c) Gametophytic (d) Both a & c

2504. The most commonly used rootstock for apple.

(a) M27 (b) M29
(c) M9 (d) M19

2505. Multiplication of apple rootstock is through.

(a) Seedling (b) **Mound layering**

(c) Trench layering (d) Tissue culture

2506. Recommended dwarfing rootstock for mango...

(a) Olour (b) Bappakai

(c) 13/1 **(d) Vellaikolamban**

2507. Technology for meadow orchard in guava was developed at.....

(a) CHES, Godhra (b**) CISH, Lucknow**

(c) IIHR, Banglore (d) BSKKV, Dapoli

2508. Technology for meadow orchard in guava was developed by......

(a) Dr.Gorakh Singh (b) Dr. H.P. Singh

(c) Dr. Giriraj Singh (d) Dr. A.K. Singh

2509. Spacing recommended for meadow orchard in guava is ....

(a) 1 x 2m **(b) 2 x 1m**

(c) 1 x 3m (d) 3 x 1m

2510. The most adopted training system in apple is.....

(a) Spindle bush (b) Central leader

(c) Modified central leader (d) Vertical axis

2511. Training system followed in pears is......

(a) Spindle bush (b) Central leader

(c) Modified central leader (d) Vertical axis

2512. The chemical used for control of vegetative growth in temperate fruits.....

(a) CCC (b) Alar

(c) Paclobutrazol (d) MCP

2513. Generally pruning can be done in apple during......

(a) Active growth stage (b**) Dormant stage**

(c) Flowering stage (d) Fruit development stage

2514. 'Dutch cut' is useful for......

(a) Wide crotch angle of branches

(b) Narrow crotch angle of branches

(c) Production of more water shoots

(d) Production of spurs

2515. 'Snaking' is a kind of.....

(a) Young tree training technique

(b) Leader management technique

(c) Type of pruning cut

(d) Method of harvesting

2516. Scaffold branching pattern can be naturally seen in.....

(a) Mango tree **(b) Pomegranate tree**

(c) Christmas tree (d) Cherry tree

2517. Cloth pins can be used to......

(a) To develop straight central leader

(b) To develop narrow crotch angle of branches

(c) To develop zigzag central leader

(d) To develop wide crotch angle of branches

2518. Pinching is commonly used to.....

(a) To induce single branch (b) To suppress branching

(c) To induce multiple branching **(d) None of these**

2519. Among the leader management techniques, bagging is used to ....

(a) Induce branching (b) Reduce branching

(c) Isolate the leader (d) Induce flowering

2520. In apple, flowering occurs on the short branches which are two or more year old, called as.....

(a) Sports (b) Suckers

(c) Scions (d) Spurs

2521. Generally, tip bearing cultivars have......

(a) Narrow crotch angle of branches

(b) Wide crotch angle of branches

(c) Branches without crotch angle

(d) None of these

2522. In general, grafting ....

(a) Induce Vigorous growth **(b) Delays maturity**

(c) Induce Dwarfing (d) None of these

2523. In general, grafting:

(a) Reduce vigour

(b) Induce early maturity of tree

(c) Induce dwarfing

(d) All of these

2524. Annual pruning is generally not desirable in:

(a) Mango (b) Sapota

(c) Jackfruit **(d) All of these**

2525. Annual pruning is generally desirable in.....

(a) Mango (b) Guava

(c) Pomegranate **(d) Phalsa**

2526. Annual pruning is recommended in.....

(a) Mango **(b) Ber**

(c) Sapota (d) None of these

2527. In guava, profuse flowering generally occurs on......

(a) Newly initiated young shoots (b) One year old shoots

(c) Two-three year old shoots (d) None of these

2528. To obtain bigger fruits, the common practice is ....

(a) Thinning (b) Bending

(c) Pruning (d) None of these

2529. Orientation of rows is most important in ....

(a) Apple (b) Pear

(c) Mango (d) Grape

2530. Wind direction is most important in ....

(a) Apple (b) Pear

(c) Mango **(d) Grape**

2531. In Maharashtra, the most widely adopted training system in grape ....

(a) Trellis system (b) Telephone system

(c) Flat roof gable system **(d) None of these**

2532. Which group of apple usually develop narrow crotch angle of branches?

(a) Spur bearer (b) Tip bearer

(c) Intermediate (d) None of these

2533. 'Spur invigoration' is a practice in which.......

(a) More spurs were forced to develop

(b) Spurs were reduced by thinning out excess spurs

(c) Branches were treated with PGR

(d) None of these

2534. The pruning method used to rejuvenate the old orchard of mango,

(a) Heading back **(b) Pinching**

(c) Dutch cut (d) None of these

2535. In meadow orchard system of guava, the plants were pruned ......

(a) Single time per year (b) Two times per year

(c) Three times per year (d) Four times per year

2536. In ber, pruning was done immediately after.....

(a) Flowering (b) New vegetative growth

(c) Fruit set **(d) Harvesting**

2537. Epicotyl or softwood grafts were preferred to develop high density orchard of mango because.......

(a) To develop single straight stem

(b) To keep desired number of branches initially

(c) To train plants from lower height

(d) All of these

2538. The fruit crop in which very less information is available on canopy management.

(a) Sapota (b) Mango

(c) Citrus **(d) Pomegranate**

2539. In apple planting material, 'whip' means.....

(a) Plant with two branches (b) Plant without roots

(c) Unbranched plants (d) None of these

2540. In apple planting material, 'feathers' means.....

(a) Unbranched plants (b) Well branched plants

(c) Plants without roots (d) Plants without buds

2541. Wide crotch angle of branches can be developed by using.....

(a) Cloth pins **(b) Tooth picks**

(c) Dutch cut (d) All of these

2542. The slender spindle training system was given by......

(a) Wertheim (b) Lespinesse

(c) Barritt **(d) None of these**

2543. 'Vertical axis' training system is also known as.........

(a) French axe (b) American axe

(c) Canadian axe (d) None of these

2544. The tree trained in slender spindle system typically has......

(a) Single permanent tier of branches

(b) Two permanent tier of branches

(c) Three permanent tier of branches

(d) No permanent tier of branches

2545. In slender spindle training system, the height of the tree was kept limited by......

(a) **Pinching**
(b) Leader renewal technique
(c) Stripping
(d) None of these

2546. The apple trees require permanent support system to each tree when trained in ......

(a) Vertical axis
(b) **Central leader**
(c) Modified central leader
(d) None of these

2547. 'Notching' can be used to promote ....

(a) **Flowering**
(b) Rooting
(c) Fruiting
(d) Branching

2548. Summer pruning in apple was practiced for the purpose......

(a) To improve flowering
(b) To improve root growth
(c) To improve fruit colour
(d) **None of these**

2549. In tall spindle system of training, there are ....

(a) One permanent tier of branches
(b) Two permanent tier of branches
(c) **No permanent tier of branches**
(d) None of these

2550. Highly feathered nursery plants are necessary for planting in case of ....

(a) Tall spindle system
(b) Vertical axis system
(c) **Slender spindle system**
(d) None of these

2551. Tip bearing apple cultivars includes......

(a) **Cortland**
(b) Rome
(c) Granny Smith
(d) All of these

2552. ________ are the foundation for successful tall spindle training system of apple.

(a) Vigorous rootstocks
(b) Dwarfing rootstocks
(c) Standard rootstocks
(d) **None of these**

2553. _______ is fire-blight resistant rootstock of apple.

(a) G-11 **(b) M-9**

(c) B-11 (d) None of these

2554. The final height of plants under tall spindle training system will be ......

(a) 6-7 ft (b) 10-11 ft

(c) 14-15 ft (d) 19-20 ft

2555. Ber is considered as 'king of arid fruit crops' due to......

(a) Deep & extensive root system

(b) Shed leaves during summer & become quiescent

(c) Hard, pubescent scales develop on buds

(d) All of these

2556. ____________ is an octaploid cultivar of ber.

(a) Illaichi (b) Umran

(c) Gola (d) None of these

2557. In ber, regular annual pruning is necessary because flowering occurs mainly on........

(a) Current season's new branches

(b) Previous season's old branches

(c) Two-three year old branches

(d) None of these

2558. In ber, annual pruning is done during.........

(a) After harvesting (b) After new leaves initiated

(c) After flowering (d) After dormant period

2559. Purple passionfruit is normally grafted on yellow passionfruit to protect plants from.......

(a) Aphids (b) Root weevil

(c) Nematodes (d) Root suckers

2560. Which rootstock is for grape?

(a) Dogridge (b) Muvandan

(c) Kurukkan (d) Pink Selection

2561. In sub-tropical agro-climatic conditions, ber tree ________ during summer.

(a) Do not shed leaves **(b) Shed leaves**

(c) Do not flowers (d) Flower

2562. M-9 is a dwarfing rootstock of......

(a) Pear (b) Plum

(c) Apple (d) Apricot

2563. In grape grown in temperate regions, pruning generally done ......

(a) One time in a year (b) Two time in a year

(c) At alternate years (d) At biennially

2564. In grape, pruning done in April month is also known as......

(a) Fruit pruning (b) Forward pruning

(c) Foundation pruning (d) None of these

2565. Forward pruning in grape is also known as

(a) Fruit pruning (b) Cane pruning

(c) Foundation pruning (d) None of these

2566. Flat roof gable system is popular among the grape growers because......

(a) Bunches are protected from sunlight but shoots are well exposed to sunlight

(b) Bunches hang within the reach of workers

(c) Spraying of chemicals on bunches & movement of farm machineries easy

(d) All of these

2567. Topping or hedging is important technique to increase yield in...

(a) Sapota (b) Grape

(c) Guava (d) Ber

2568. Growth of guava trees can be checked by.......

(a) Root pruning (b) Water stress

(c) Cold **(d) All of these**

2569. Dwarfing rootstock developed for guava at IARI, New Delhi

(a) Dogridge **(b) Aneuploid no. 82**

(c) Vellaikolamban (d) *Psidium indica*

2570. In guava, ______________ bahar is taken to obtain high quality fruits.

(a) Ambebahar **(b) Mrigbahar**

(c) Hasthbahar (d) None of these

2571. The fruits from ambebahar generally ripen during ......

(a) Jan-Feb (b) April-May

(c) July-August (d) Oct-Nov

2572. ____________ practice is adopted to regulate crop for guava grown on heavy soils.

(a) Withholding irrigation (b) Application of PGR

(c) Root pruning (d) None of these

2573. ____________ practice is adopted to regulate crop for guava grown on lighter soils.

(a) Withholding irrigation (b) Application of PGR

(c) Root pruning (d) None of these

2574. In guava, ____________ crop cycles per year are possible.

(a) One **(b) Two**

(c) Three (d) Four

2575. In guava, pruning generally done after ......

(a) Flowering (b) New leaf initiation

(c) Harvesting (d) None

2576. The tree in which top portion grows vigorously which suppress the growth of lower side branches, termed as ....

(a) Basitonic (b) Gravitonic

(c) Plagiotonic **(d) Acrotonic**

2577. The tree in which lower branches grows vigorously which suppress the growth of central leader, termed as ....

(a) Basitonic (b) Gravitonic

(c) Plagiotonic (d) Acrotonic

2578. The trees with _______________ growth habit are easier to manage in conical shape of canopy.

(a) Basitonic (b) Gravitonic

(c) Plagiotonic (d) Acrotonic

2579. Pome fruit trees (apple, pear) produce fruits on the branch which is _______________.

(a) One year old (b) **Three year old**

(c) Current year growth (d) Six month old

2580. Stone fruit trees (plum, peach, cherry) produce fruits on the branch which is _______________.

(a) One year old (b) **Two year old**

(c) Current year growth (d) Six month old

2581. Through orientation of branch by controlling the flow of _______________, the branches can be forced to produce either vegetative growth or flowers.

(a) Cytokinin **(b) Auxin**

(c) Gibberelin (d) None of these

2582. Which variety is recommended by IARI for high density planting in mango?

(a) Alphonso (b) Sonpari

(c) Dashehari **(d) Amrapali**

2583. For induction of flowers in biennial mango cultivars, the commonly used chemical is ......

(a) Potassium nitrate (b) Ethrel

(c) Paclobutrazol (d) NAA

2584. Which salt tolerant rootstock has been used at Reliance Jamnagar mango orchard?

(a) Vellaikolamban **(b) 13/1**

(c) Saltcreek (d) Olour

2585. ____________ is a vigorous variety of mango.

(a) Amrapali (b) Neeleshwari

(c) MadrasiHafus **(d) Alphonso**

2586. ____________ is a dwarf variety of mango.

(a) Alphonso (b) Kesar

(c) Sonpari **(d) Amrapali**

2587. Which variety of mango was extensively used as parent to induce regularity in bearing in mango breeding?

(a) Totapuri **(b) Neelam**

(c) Dashehari (d) Alphonso

2588. In India, the reason for low productivity in mango includes.....

(a) Poor canopy management (b) Old/senile orchards

(c) No use of dwarfing rootstock **(d) All of these**

2589. In papaya, which variety is recommended for high density planting?

(a) Red Lady (b) Co-1

(c) Pusa Nanha (d) None of these

2590. With Pusa Nanha variety of papaya, accommodation of ________ plants/hectare is possible.

(a) 4433 **(b) 6400**

(c) 8033 (d) None of these

2591. In pineapple high-density planting, accommodation of __________ plants per hectare is possible.

(a) 33000 (b) 23000
(c) 63000 (d) 83000

2592. Fig trees are generally trained in __________ system.

(a) Central leader **(b) Vase**
(c) Slender Spindle (d) None of these

2593. In pomegranate, the type of training system generally adopted in India is....

(a) Multi-stem plants (b) Two stem plants
(c) Single stem plants (d) None of these

2594. In grape, __________ is considered as essential operation to obtain yield.

(a) Spraying of PGR (b) Berry thinning
(c) Pruning (d) None of these

2595. __________ plants per hectare can be accommodated in meadow orchard system of guava.

(a) 5000 (b) 3000
(c) 6000 (d) 2000

2596. The height of the properly managed plants of guava under meadow orchard should be....

(a) 2.0-2.5m **(b) 1.0-1.5m**
(c) 0.5m (d) None of these

2597. In meadow orchard system, pruning will be done ....

(a) One time in a year. (b) Two times in a year
(c) Three times in a year (d) Four times in a year

2598. In Cashew, which system of training is generally adopted?

(a) Modified Central leader (b) Central leader
(c) Slender spindle (d) Spindle bush

2599. In fig, pruning generally done ....

(a) **Every year** (b) Every two year

(c) Every three year (d) None of these

2600. In ber, pruning is done in ....

(a) Winter months (b) **Summer months**

(c) Monsoon (d) None of these

2601. For high density planting in mango cv. Amrapali, the recommended spacing is.....

(a) 3.5x3.0m (b) **2.5 x 2.5m**

(c) 2x1.5m (d) 3.5x1.5m

2602. ________ is a dwarf variety of sapota.

(a) PKM-1 (b) PKM-2

(c) **PKM-3** (d) None of these

2603. For cherry, ________ is a dwarfing rootstock.

(a) Quince (b) Pixy

(c) **Colt** (d) Bud-9

2604. Most of the apple varieties from Delicious group have ____________ bearing habit.

(a) Intermediate (b) Tip

(c) **Spur** (d) None of these

2605. In general, apple varieties with spur bearing habit have ____________ growth habit.

(a) **Spreading** (b) Upright

(c) Intermediate (d) None of these

2606. In general, tip bearing apple cultivars produce ____________.

(a) **Few or no spurs** (b) High number of spurs

(c) Intermediate number of spurs (d) None of these

2607. On the basis of growth & bearing habit in apple,________ variety is grouped under intermediate bearer.

(a) **Honeycrisp** (b) Golden Delicious

(c) Cortland (d) None of these

2608. Most of the apple varieties from Delicious group have ____________ growth habit.

(a) Acrotonic (b) Intermediate

(c) **Basitonic** (d) None of these

2609. Hybrid tree cone (HYTEC) tree training system of apple was given by______________.

(a) **Barritt** (b) Lespinasse

(c) Wertheim (d) None of these

2610. In apple rootstock name with 'M' prefix stands for ____________.

(a) Michigan Apple Clone (b) Merton

(c) **East Malling** (d) None of these

2611. In apple, major purpose of summer pruning is...........

(a) To reduce canopy volume (b) To improve fruit colour

(c) **Both 'a' and 'b'** (d) None of these

2612. Summer pruning is generally beneficial to ________ group of apple varieties.

(a) Tip bearer (b) **Spur bearer**

(c) Intermediate (d) None of these

2613. Summer pruning in apple should be done ________________

(a) About 3-6 months before harvest

(b) About 3-6 weeks after harvest

(c) About 3-6 months after harvest

(d) **About 3-6 weeks before harvest**

2614. National fruit crop of India is

(a) **Mango** (b) Banana

(c) Citrus (d) Papaya

2615. Refractrometer is used for measuring

(a) Acid (b) **T.S.S**

(c) Salt (d) Fiber

2616. In fruit and vegetable production, India rank .... position in the world.

(a) 2nd (b) 1st

(c) 5th (d) 3rd

2617. Sonpari variety of mango is released by......

(a) NAU (b) AAU

(c) JAU (d) SDAU

2618. ________ is used in mango for promoting the early flowering.

(a) Paclobutrazol (b) Gibberellins

(c) Cytokinin (d) Auxin

2619. ______ is the latest method of irrigation and is a more specialized one.

(a) Porous **(b) Drip**

(c) Flood (d) All the above

2620. Balanagar is the famous variety of

(a) Ber **(b) Custard Apple**

(c) Guava (d) Aonla

2621. ________ is a ripening hormone.

(a) Ethylene (b) Giberellin

(c) Cytokinin (d) None

2622. Coconut Research Station of Gujarat is located at ________.

(a) Mahuva (b) Navsari

(c) Junagadh (d) Anand

2623. Objective of pre-cooling is

(a) To remove field heat (b) To fresh the product

(c) To improve the quality (d) None of these

2624. Pectin is measured by

(a) Thermometer (b) Refractrometer

(c) Jelly meter (d) All of above

2625. Fruit which is used for preparation of good quality jelly is

(a) Mango (b) Banana

(c) Guava (d) Sapota

2626. Solution of salt water is called as

(a) Brine (b) Vinegar

(c) Juice (d) Cider

2627. Token money means

(a) Gold & silver **(b) Cheaper metal**

(c) Large coins (d) All the above

2628. World forest day is celebrated on every

(a) 5th June (b) 3rd October

(c) 21st March (d) 14th February

2629. Institute of Arid Zone Forestry Research is located in ______.

(a) New Delhi **(b) Jodhpur**

(c) Dehra Dun (d) Bangalore

2630. Urea contain ____ % N

(a) 42 (b) 44

(c) 46 (d) 48

2631. Date palm Research station in Gujarat is located at

(a) Mundra (b) Mahuva

(c) Jagudan (d) Godhra

2632. Father of economics

(a) Adam Smith (b) Marshall

(c) Robbins (d) None of these

2633. World environment day is celebrated on every

(a) **5th June** (b) 3rd October
(c) 11th July (d) 31st December

2634. Main Regional station of Spices in Gujarat is located at

(a) Navsari (b) Anand
(c) Patan (d) **Jagudan**

2635. Application of fertilizer with drip irrigation is known as

(a) Foliar spray (b) Sprinkler irrigation
(c) **Fertigation** (d) Surface irrigation

2636. Alternate bearing is commonly found in

(a) Sapota (b) Citrus
(c) Guava (d) **Mango**

2637. Brix degree is an unit of measurement of

(a) Salt content in juice (b) Sucrose content in juice
(c) **Total solids in juice** (d) Juice content of fruit

2638. Papain is prepared from

(a) Mango (b) Sapota
(c) Banana (d) **Papaya**

2639. Epicotyl grafting is done in mango when seedlings are ______days old.

(a) 10 (b) **15**
(c) 20 (d) 25

2640. India is the largest producer and exporter of

(a) **Mango** (b) Sapota
(c) Banana (d) Grapes

2641. IBA is widely used for

(a) **Rooting of cutting** (b) Breaking dormancy
(c) Fruit setting (d) Flower drop

2642. In Air-layering method ________ is commercially used on the cut surface.

(a) **Sphagnum moss**
(b) Perlite
(c) Vermiculite
(d) Sand

2643. Which of the following is non-selective herbicides

(a) 2,4-D
(b) Fluchloralin
(c) **Glyphosate**
(d) None of these

2644. Solution of sugar in water is called as

(a) Jelly
(b) **Syrup**
(c) Jam
(d) Marmalade

2645. Example of Copper fungicide is

(a) Thiram
(b) Captafol
(c) Carbendazim
(d) **Bordeaux mixture**

2646. The cold treatment (0-5°C) is known as

(a) Scarification
(b) **Stratification**
(c) Mechanical treatment
(d) None of above

2647. ________ variety of mango is free from spongy tissue.

(a) Alphonso
(b) Sindhu
(c) **Ratna**
(d) Kesar

2648. Jam is prepared from

(a) Fruit peel
(b) Whole fruit
(c) **Fruit pulp**
(d) None of above

2649. Kesar variety of mango is the origin of

(a) Uttar Pradesh
(b) Rajasthan
(c) Maharashtra
(d) **Gujarat**

2650. _____ is the development of fruit without fertilization of the ovule

(a) Pollination
(b) Hybridization
(c) Transformation
(d) **Parthenocarpy**

2651. Tapka is associated with maturity of

(a) **Mango** (b) Sapota

(c) Papaya (d) Grapes

2652. CO_2 concentration inside the green house is

(a) 300 ppm (b) 400 ppm

(c) 250 ppm (d) **1000 ppm**

2653. Removal of outer cover of fruit is known as

(a) Lying (b) **Peeling**

(c) Cutting (d) All the above

2654. Sapota is commercially propagated on __________ rootstock.

(a) Mango (b) **Rayan**

(c) Sapota (d) None

2655. Parthenocarpic fruits are

(a) False fruits (b) Composite fruits

(c) **Seedless fruits** (d) None of these

2656. __________ variety of mango is susceptible to spongy tissue.

(a) Kesar (b) **Alphonso**

(c) Ratna (d) Dashehari

2657. Ber is the __________ of arid fruits

(a) Master (b) Queen

(c) **King** (d) Minor

2658. __________ is used for synchronize the flowering in pineapple

(a) **NAA** (b) IBA

(c) IAA (d) GA3

2659. __________ is useful propagation unit for rooting of cutting.

(a) **Mist Chamber** (b) Green House

(c) Net House (d) Polyhouse

2660. Green manure crop may be turned in soil at the stage of

(a) Harvesting (b) Branching

(c) 50% Flowering (d) None of these

2661. Mango variety which is suitable for high density planting

(a) Amrapali (b) Kesar

(c) Alphonso (d) Rajapuri

2662. Now a day's banana is commercially propagated by

(a) Seed **(b) Tissue culture**

(c) Grafting (d) Layering

2663. ________ is known as king of spices.

(a) Black pepper (b) Cardamom

(c) Cinnamon (d) Clove

2664. Which preservative is a source of benzoic acid

(a) KMS (b) Citric acid

(c) Sodium benzoate (d) Acetic acid

2665. Common name of Paclobutrazol in market is________

(a) Cultar (b) Atrazine

(c) Trizole (d) Bultar

2666. ________is also known as Kalpavriksha.

(a) Coconut (b) Areca nut

(c) Cashew nut (d) Banana

2667. An ideal fruit for making jelly should be rich in

(a) Acid and proteins **(b) Pectin and acids**

(c) Pectin and sugars (d) Sugar and acids

2668. Which variety of cashew is suitable for commercial cultivation in Gujarat?

(a) Ullal – 3 **(b) Vengurla-4**

(c) Ankayyam -1 (d) Priyanka

2669. Mango pulp is preserved by

(a) **KMS** (b) Salt

(c) Sugar (d) Citric acid

2670. Deficiency of Vitamin A causes

(a) **Night blindness** (b) Beriberi

(c) Scurvy (d) Common cold

2671. __________ is the practice of covering soil around plants

(a) **Mulching** (b) Fumigation

(c) Solarisation (d) None of these

2672. Removal of apical bud is called

(a) Pruning (b) Cutting

(c) **Pinching** (d) Bending

2673. Richest source of vitamin C is

(a) Orange (b) **Aonla**

(c) Bael (d) Citrus

2674. Fenny is prepared by fermentation of

(a) Coconut (b) Apple

(c) Grapes (d) **Cashew**

2675. Epicotyl grafting is followed in __________ fruit crop

(a) Grapes (b) Citrus

(c) **Mango** (d) Sapota

2676. Thickness of UV stabilized film for poly house should be

(a) 500 micron (b) **200 micron**

(c) 50 micron (d) 100 micron

2677. Nutrients which are required by plants in small quantities are referred as

(a) Macro nutrients (b) **Micronutrients**

(c) Primary nutrients (d) All the above

2678. Two or more crops are grown together on the same piece of land the in row is known as

(a) Intra cropping
(b) Succession cropping
(c) Intercropping
(d) None of these

2679. Deficiency of Vitamin B_1 Cause disease of

(a) Night blindness
(b) Beri-Beri
(c) Goitre
(d) Weak bones

2680. 'Cricket Ball' is a popular variety of

(a) Pomegranate
(b) Mango
(c) Citrus
(d) Sapota

2681. The Word horticulture is derived from 'Hortus' and 'Cultura' meaning Garden and Cultivation. These words are ________.

(a) Latin
(b) Greek
(c) Persian
(d) English

2682. ___________System accommodates 16 % more tree per hectare as compared to square system.

(a) Hexagonal
(b) Triangular
(c) Rectangular
(d) Diagonal

2683. ___________ is the practice of covering soil around plants.

(a) Mulching
(b) Fumigation
(c) Solarization
(d) Ridging

2684. ___________ is a non selective herbicide.

(a) Glyphosate
(b) Dalapon
(c) 2-4-D
(d) Fluchloralin

2685. _________Plant growth regulator is used as weedicide/Herbicide.

(a) Auxin
(b) Gibberellin
(c) Cytokinin
(d) Ethylene

2686. Highest ethylene is produced in _________ fruit crop.

(a) Apple (b) Mango

(c) Banana (d) Sapota

2687. _________ is a highly salt tolerant crop

(a) Datepalm (b) Mango

(c) Banana (d) Citrus

2688. Which of the following is poor man's fruit?

(a) Ber (b) Mango

(c) Grapes (d) Apple

2689. In Gujarat there are _________ types of agro-climatic zones.

(a) 7 **(b) 8**

(c) 9 (d) 10

2690. Pruning is a _________ process.

(a) Talling **(b) Dwarfing**

(c) Shortening (d) All of above

2691. The recommended spacing for banana is _________.

(a) 1.8x1.8m (b) 2x2m

(c) 3x3m (d) 4x4m

2692. The thermal death point of most plant cell is around _________.

(a) 30 °C (b) 40 °C

(c) 60 °C (d) 70 °C

2693. Horticultural crops are rich source of vitamins and minerals so they are called _________

(a) Protective food (b) Staple food

(c) Sustainable food (d) None

2694. _________ is a branch which deals with the scientific study of fruits.

(a) Olericulture (b) Floriculture

(c) Tissue culture **(d) Pomology**

2695. Seed of pomegranate is called ________

(a) Nut **(b) Aril**

(c) Drupe (d) Stone

2696. Botanically, Sapota fruit is known as ________.

(a) Pome (b) Drupe

(c) Berry (d) Nut

2697. Parthenocarpic fruits are ________

(a) False fruit (b) Composite fruit

(c) Multiple fruits **(d) Seedless fruits**

2698. ________ is a non-climacteric fruit.

(a) Banana **(b) Pomegranate**

(c) Sapota (d) Guava

2699. Best soil for planting a fruit orchard should have pH between ________

(a) 3.5-5.5 **(b) 6.5-7.5**

(c) 7.5-10.5 (d) 4.5-6.5

2700. National fruit of India is________.

(a) Apple (b) Papaya

(c) Banana **(d) Mango**

2701. ________ is also called as a friend of farmers.

(a) Earthworm (b) Silkworm

(c) Bacteria (d) Rat

2702. ________ is done to give size and shape to plants.

(a) Pruning **(b) Training**

(c) Thinning (d) None of above

2703. There are ________elements have been considered as essential for higher green plant.

(a) 14 (b) 17

(c) 15 **(d) 16**

2704. __________ is a nitrogen fixing bacteria.

(a) Azotobacter (b) Phosphobacter

(c) Bacillus (d) Aspergillus

2705. __________ is a highly acid tolerant crop.

(a) Bael (b) Pomegranate

(c) Jamun (d) Phalsa

2706. Botanically pomegranate is __________ type of fruit.

(a) Balausta (b) Pepo

(c) Drupe (d) Berry

2707. Water requirement of seed is __________ times more than its weight.

(a) 2 to 3 (b) 3 to 4

(c) 1 to 2 (d) 4 to 5

2708. In green house carbon dioxide concentration is __________ PPM.

(a) 800 to 900 (b) 800 to 1000

(c) 800 to 1100 **(d) 800 to 1200**

2709. Raising of plants by seed is called __________.

(a) Sexual propagation (b) Asexual propagation

(c) Tissue culture (d) None of the above

2710. ________is temperate region fruit.

(a) Mango (b) Sapota

(c) Banana **(d) Apple**

2711. Aonla is mainly propagated by __________.

(a) Budding (b) Grafting

(c) Tissue culture (d) Seed

2712. Ginger and Turmeric is propagated by __________.

(a) Rhizome (b) Bulb

(c) Runner (d) Suckers

2713. Lalit is improved variety of ________.

(a) Ber (b) **Guava**

(c) Papaya (d) Apple

2714. Guava bears mostly on ________

(a) Current seasons growth (b) One year old shoot

(c) Spurs (d) Very old shoot

2715. ____________ chemical is used for breaking of seed dormancy

(a) Thiourea (b) IAA

(c) IBA (d) NAA

2716. ____________ is a mature ovary with associated parts

(a) Fruit (b) Flower

(c) Leaf (d) Stem

2717. The term __________ crops refers to those perennial crop which are cultivated on an extensive scale in a large continuous area, owned & managed by an individual or a company.

(a) Plantation (b) Tropical

(c) Temperate (d) Vegetable

2718. Sugandham is famous variety of

(a) Black piper (b) Cinnamon

(c) Cardamom **(d) Turmeric**

2719. Neelphonso, Neeleshwari, Neeleshan variety of mango is released by __________

(a) NAU (b) AAU

(c) JAU (d) SDAU

2720. Grand Naine is a leading variety of ________.

(a) Banana (b) Mango

(c) Sapota (d) Papaya

2721. In Guava fruits which bahar is recommended __________.

(a) Ambe bahar (b) Hasth bahar

(c) Mrig bahar (d) All of the above

2722. __________ is a basic plant nutrient

(a) Nitrogen (b) Phosphorous

(c) Carbon (d) Sulphur

2723. Saline soil is reclaimed by __________

(a) Gypsum **(b) Liming**

(c) leaching (d) None

2724. The cultivation and management of forest tree is known as__________.

(a) Horticulture (b) Agronomy

(c) Agro forestry (d) Silviculture

2725. Arid region receive ....mm average annual rainfall

(a) <250 (b) >250

(c) 250-500 (d) >500

2726. Example of variable cost

(a) Machinery & equipment **(b) Seed & fertilizer**

(c) Insurance premium (d) None of these

2727. "Deep ploughing during summer to expose pupa against sunlight or predator" is a ...... method of insect control

(a) Natural control **(b) Cultural**

(c) Physical (d) Mechanical

2728. __________ is the progeny of breeder seed.

(a) Certified seed (b) Nucleus seed

(c) Registered seed **(d) Foundation seed**

2729. Coming in contact of a pathogen with a host plant is called.....

(a) Infection (b) Penetration

(c) Inoculation (d) Infestation

2730. For storage moisture content of seed should not exceed more than ______%

(a) 20 (b) 10

(c) 25 **(d) 12**

2731. Which is colorful and often scented part of the flower that attracts insects

(a) Petal (b) Sepal

(c) Anther (d) All above

2732. Disease may be endemic in one region and epidemic in another, which prevalent throughout country is called

(a) Sporadic disease **(b) Pandemic disease**

(c) Endemic disease (d) Epidemic disease

2733. ________a fertilized mature ovule consisting of embryo storage material and protective coats.

(a) Seed (b) Bud

(c) Fruit (d) None of this

2734. Any agent that cause damage; include all living organism associated with the disease viruses as well as the abiotic agents such as adverse atmosphere or soil conditions.

(a) Biotrophs (b) Parasites

(c) Saptophytes **(d) Pathogen**

2735. Dieback, Root rot, Damping off, Scab *etc* are the example of....

(a) Virus (b) Bacteria

(c) Fungi (d) MLO's

2736. Trueness to types varietals purity plants/seeds conforming to characteristics of the variety as described by the breeder

(a) Physical purity (b) Chemical purity

(c) Genetic purity (d) None of this

2737. Golden yellow colored certificate tag is issued for class of seed

(a) Certified seed (b) Nucleus seed

(c) Registered seed **(d) Breeder seed**

2738. Partial or semi-stem parasite found on Mango crop is ______

(a) *Cuscuta* sp. (b) *Orobanche* sp.

(c) *Loranthus* sp. **(d) *Striga* sp.**

2739. Mancozeb is an example of ______

(a) Antibiotics (b) Eradicant

(c) Nematicide **(d) Protectant**

2740. Which of the following vitamin is known as an antioxidant ______

(a) Vit. C (b) Vit. A

(c) Vit. E (d) Vit. K

2741. What should be the true direction of green house in field condition?

(a) East- West **(b) North- South**

(c) West- North (d) North- West

2742. Transfer of pollen grain from anther to stigma is called as ______.

(a) Pollination (b) Hybridization

(c) Parthenocarpy (d) Fertilization

2743. Which preservative is most suitable for colored juice

(a) KMS (b) **Sodium Benzoate**

(c) Hydrogen peroxide (d) Ascorbic acid

2744. Solution of sugar in water is called as ______

(a) Jelly (b) Vinegar

(c) Juice **(d) Syrup**

2745. Refractrometer is used for measuring ______

(a) Salt % (b) Acid %

(c) Sugar % (d) Fiber %

2746. Cobalt 60 is the source for

(a) **Irradiation** (b) Sterilization

(c) Pasteurization (d) Balancing

2747. Post harvest losses of fruit and vegetable in India is about ________.

(a) 10 to 15% (b) 20%

(c) 25 to 30% (d) 50%

2748. Repetition of treatment is known as ________

(a) Replication (b) Randomization

(c) Distribution (d) None of these

2749. Most heavily pruned fruit tree is

(a) *Syzygium cuminii* (b) *Ficus carica*

(c) *Grewia subinequalis* (d) *Morus alba*

2750. Botanical name of ramphal (Bullock's heart)

(a) *A. glabra* **(b) *A. reticulata***

(c) *A. squamosa* (d) *A. cherimola*

2751. African Pride, annona variety is a cross between..........

(a) Sugar apple & *Annona cherimoya*

(b) Sharifa & *Annona cherimoya*

(c) Custard apple & *Annona cherimoya*

(d) All of the above

2752. The botanical name of wood apple

(a) *Feronia marmelos* (b) *Lemonia marmelos*

(c) *Feronia limonia* (d) None of the above

2753. Fruit type of wood apple and Bael

(a) Hespiridium (b) Balausta

(c) Amphisaraca (d) Drupe

2754. Arka Sahan is a hybrid of annona developed as a result of cross between

(a) *Annona reticulata* X *Annona squamosa*

(b) *Annona atemoya* X *Annona squamosa*

(c) *Annona squamosa* X *Annona glabra*

(d) None

2755. Chromosome number of custard apple (2n)

(a) 7 (b) 12

(c) 14 (d) 16

2756. In some countries, pond apple tree is also considered as........

(a) Climbers (b) Hedge

(c) Shrub **(d) Weed**

2757. Native place of Karonda is same as that of

(a) Jamun (b) Cordia

(c) Phalsa **(d) All**

2758. Bullock's heart is used as rootstock in

(a) Custard apple (b) Fig

(c) Guava (d) Aonla

2759. Which is the most favourite annonaceous fruit in India

(a) Ramphal **(b) Sitaphal**

(c) Lakshamanphal (d) Hanumanphal

2760. Red Sitaphal is a variety of annona evolved as......

(a) Mutation (b) Introduction

(c) Hybridization **(d) Chance seedling**

2761. In English, Wood apple is also known as........

(a) Elephant fruit **(b) Elephant apple**

(c) Apple of arid (d) Hardy apple

2762. Wood apple is native to ....

(a) Japan **(b) India**

(c) Srilanka (d) Maxico

2763. Which fruit has highest iron content

(a) Dry Karonda (b) Green Mango

(c) Walnut (d) Date Palm

2764. Which of the following is/ are the rootstock of custard apple Pond apple b) Pala c) Ramphal

(a) a and b (b) b and c

(c) a and c (d) None

2765. Sugar content of custard apple %

(a) 45 (b) 15

(c) 12 **(d) 20**

2766. Highest ascorbic acid content is present in ....

(a) Barbados Cherry (b) Citrus

(c) Aonla (d) Guava

2767. The somatic chromosome number of *Syzygium cuminii*

(a) 14 (b) 18

(c) 40 (d) 36

2768. The richest source of riboflavin is present in.............

(a) Bael (b) Papaya

(c) Pineapple (d) Cashew Nut

2769. The highest Ca content is present in...........

(a) Dry Karonda **(b) Litchi**

(c) Woodapple (d) Green Mango

2770. Name the fruit crop which is used as hedge...........

(a) Duranta (b) Kikkar

(c) Karonda (d) Tecoma

2771. For arid climatic condition which fruit should not be selected?

(a) Date Palm
(b) Ber
(c) Pomegranate
(d) Guava

2772. Date palm should be propagated by ....

(a) Runners
(b) Suckers
(c) Seed
(d) Rhizomes

2773. Aonla is also called as

(a) Fruit of the 21st centuary
(b) Amritphal fruit
(c) Indian goose berry
(d) All of the above

2774. Edible part of pomegranate is ....

(a) Juicy Seed Coat
(b) Pericarp
(c) Fleshy Peduncle
(d) Fleshy Receptacle

2775. Cherimoya is mostly restricted to.........

(a) Assam and Hills of South India
(b) Gujarat and Maharshtra
(c) Jammu and Kashmir
(d) Kerala and Tamil Nadu

2776. Active compound present in bael fruit is.............

(a) Marmelosin
(b) Annonine
(c) Ficin
(d) Bromelin

2777. Balanagar is variety of

(a) Ber
(b) Pomegranate
(c) Custard apple
(d) Fig

2778. *Carissa grandiflora* (Natal plum) is originated to ....

(a) Australia
(b) Maxico
(c) Japan
(d) South Africa

2779. The term five corner fruit is associated with............

(a) Cocoa (b) Kiwi

(c) Carambola (d) Ber

2780. Date palm contains 10.6% iron where as Karonda content ....

(a) 50% **(b) 39.1%**

(c) 30% (d) 28.5%

2781. Chirmi, Rutab, Tamer are the development stages of ....

(a) Avocado (b) Ber

(c) Date Palm (d) Bael

2782. Drink of Date Palm is known as..........

(a) Perry (b) Cidar

(c) Dibbis (d) NOTA

2783. Arrak is liquor prepared from ....

(a) Date Palm (b) Avocado

(c) a) and b) (d) Pear

2784. Date palm variety suitable for juice...

(a) Hayani (b) Barhee

(c) Halawy (d) Sevi

2785. Storage temperature of ber...............

(a) 3 ^{0}C (b) 12 ^{0}C

(c) 5 ^{0}C (d) 13 ^{0}C

2786. Barbados cherry belongs to family ....

(a) Rosaceae (b) Rutaceae

(c) Musaceae **(d) Malphigiaceae**

2787. Which of the following is/are multiple fruit...........

(a) Fig (b) Banana

(c) Mango (d) Apple, Pear

2788. Highly shade tolerant fruit crop is ....

(a) Kiwi (b) Rambutan

(c) Durian **(d) Carambola**

2789. Origin of custard apple...............

(a) S.E. Asia **(b) Tropical America**

(c) China (d) India

2790. Bael is commercially propagated through...

(a) Grafting (b) Cutting

(c) Budding (d) Seed

2791. In phalsa pruning is done in ....

(a) September (b) October

(c) November **(d) December**

2792. Kagzi, Etwah, Kagzi Gonda, Kagzi Banarasi are variety of...........

(a) Ber (b) Fig

(c) Bael (d) Loquat

2793. First fruit crop known to mankind...........

(a) Mango (b) Karonda

(c) Date Palm (d) Ber

2794. Type of inflorescence in pomegranate...............

(a) Spadix (b) Panicle

(c) Balausta **(d) Hypanthodium**

2795. *Ficus carica* is................

(a) Dioecious (b) Monoecious

(c) Gynodioecious (d) Andromonoecious

2796. Largest annona fruit in terms of weight and size is...

(a) *Annona squamosa* (b) *Annona reticulate*

(c) *Annona glabra* **(d) *Annona muricata***

2797. *Annona atemoya* bears fruit in ....

(a) Dec-Jan (b) Jan- Feb

(c) March – April (d) Sep - Oct

2798. For getting optimum yield pruning is necessary in

(a) Mango (b) Guava

(c) Phalsa (d) Annona

2799. In fig which variety is suitable for canning purpose............

(a) Conardia **(b) Excel**

(c) Deanna Fig (d) None

2800. Date palm variety which can be eaten as row............

(a) Barhee (b) Khalas

(c) Halawy **(d) All of above**

2801. Family of carambola............

(a) Rhamnaceae (b) Moraceae

(c) Oxalidaceae (d) Annonaceae

2802. Notching is commonly practiced in............

(a) Grape (b) Date Palm

(c) Fig (d) Ber

2803. Chromosome number of bael............

(a) 18 (b) 48

(c) 36 (d) 24

2804. In Subtropics Aonla exhibits ....

(a) Evergreen nature **(b) Deciduous nature**

(c) None (d) a) and b)

2805. Native of date palm............

(a) Iran (b) U.S.A.

(c) T. America **(d) Iraq**

2806. Chinese date is common name of..............

(a) Phalsa (b) Aonla

(c) Jamun **(d) Ber**

2807. Breba is parthenocarpically produced crop of..............

(a) Fig (b) Mulberry

(c) Ber (d) Aonla

2808. A fruit called Indian gooseberry is ....

(a) Aonla (b) Ber

(c) Guava (d) Date Palm

2809. Ber is propagated by............

(a) Veneer grafting (b) Seed

(c) T-budding (d) None

2810. Date fruits for fresh eating are harvested at which stage?

(a) Doka **(b) Dang**

(c) Pind (d) Gandora

2811. One Kg ripe date palm equals to

(a) 3240 calories (b) 3750 calories

(c) 3260 calories **(d) 3150 calories**

2812. Family of phalsa...........

(a) Moraceae **(b) Tilliaceae**

(c) Rutaceae (d) Proteaceae

2813. Which of the following used as substitute for tamarind...............

(a) Aonla **(b) Carambola**

(c) Mulberry (d) Avocado

2814. Acidic nature star fruit is due to ....

(a) Oxalic acid (b) Malic acid

(c) Linolic acid (d) Propenoic acid

2815. The chromosome number of Umran ber is..............

(a) 2n = 96 (b) 2n = 24

(c) 2n = 48 (d) 2n = 24=48

2816. Highly salt tolerant fruit crop is...............

(a) Date palm (b) Ber

(c) Aonla **(d) All of above**

2817. Which of the following fruit crop has highest somatic chromosome Number.....

(a) Custard apple (b) Bael

(c) Mulberry (d) Ber

2818. On the basis of vit. C content present in fruit which statement is true?

(a) Aonla> Guava>Citrus>Chilli

(b) Barbados cherry>Aonla>Guava>Chilli

(c) Barbados cherry>Guava>Aonla>Chilli

(d) Aonla>Barbados Cherry>Guava>Citrus

2819. CAZRI is lacated at ....

(a) Bikaner **(b) Jodhpur**

(c) Srinagar (d) Pune

2820. Pond apple is tolerant to flood condition which is used as a root stock..

(a) *A. squamosa* (b) *Morus alba*

(c) *Aegle marmelos* (d) *Z. mauritiana*

2821. Cauliflorous bering habit observed in ....

(a) Jackfruit (b) Star fruit

(c) Cocoa **(d) All of above**

2822. *Averrhoea bilimbi* & *Averrhoea carambola* belongs to family .....

(a) Anacardiaceae **(b) Oxalidaceae**

(c) Caricaceae (d) Rutaceae

2823. Thrum type of hetrostyly is observed in..........

(a) Almond (b) Carambola

(c) Pomegranate **(d) (a) & (b)**

2824. Insect responsible for caprification..............

(a) *Blastophega psens* (b) *Indarbela spp.*

(c) Fruit fly (d) Butter fly

2825. Type of inflorescence in fig............

(a) Hesperidium **(b) Hypanthodium**

(c) Hypanthium (d) None

2826. .... is derived from *Ficus carica x Ficus glomerata*

(a) San pedro (b) Smyrna

(c) Pune fig (d) Capri fig

2827. In India, phalsa is known as ....

(a) Star fruit **(b) Star apple**

(c) Star tree (d) None

2828. Variety of Phalsa is ....

(a) Maroon **(b) Sharabati**

(c) Sai Sharabati (d) Goma Kirti

2829. King of arid fruit is ...

(a) Bael **(b) Ber**

(c) Aonla (d) Date

2830. Which ber variety is released from CIAH, Bikaner, Rajasthan

(a) Thar Sevika (b) Sanaur -2

(c) Gola (d) None

2831. Maturing time of ber after flowering

(a) 25-50 days (b) 50-100 days

(c) 100-150 days **(d) 150-175 days**

2832. Necrosis is a physiological disorder of

(a) Annona **(b) Aonla**
(c) Phalsa (d) Jamun

2833. Centrifugal process is developed for separation of gritty portion from pulp

(a) Carambola **(b) Annona**
(c) Aonla (d) Ber

2834. Stone fruit is a physiological disorder of.......

(a) Date (b) Mulberry
(c) Carambola **(d) Custard apple**

2835. Leading state in India for pomegranate cultivation is........

(a) Tamilnadu **(b) Maharashtra**
(c) Kerala (d) Gujarat

2836. CHES is located at

(a) Godhra, Gujarat (b) Pune, Maharashtra
(c) Kota, Rajasthan (d) None

2837. NRC for pomegranate is situated at..........

(a) Solan **(b) Solapur**
(c) Muzaffarpur (d) Kolhapur

2838. Pomegranate flower is.........

(a) Male (b) Hermaphrodite
(c) Intermediate **(d) All**

2839. Among following which is the variety of Bael

(a) Goma Kirti (b) Goma Prateek
(c) Goma Yashi (d) Goma Priyanka

2840. Seedlessness in pomegranate is due to

(a) Heterostyly
(b) Lack of lignification of testa
(c) Fruit cracking
(d) None

2841. Which bahar treatment is followed in pomegranate

(a) Mrig **(b) Hasth**

(c) Ambe (d) None

2842. Which type of pomegranate is used for ornamental purpose

(a) Double flower type (b) Nana/ Dwarf type

(c) a & b both (d) None

2843. Pomegranate is ...

(a) Climacteric fruit **(b) Non Climacteric fruit**

(c) a & b (d) None

2844. Tolerant / Resistant variety of pomegranate to fruit cracking...

(a) Francis **(b) Banarasi**

(c) Dholka (d) Jyoti

2845. Internal brakdown disorder of pomegranate is seen in ... bahar

(a) Mrig bahar **(b) Ambe bahar**

(c) Hasth (d) None

2846. Among the following which variety of jamun is released from CHES

(a) Goma Khatta **(b) Goma Priyanka**

(c) Goma Prateek (d) Goma Aishwariya

2847. Among the following which variety of Cashewnut is

(a) Goma Khatta **(b) Goma Priyanka**

(c) Goma Prateek (d) Goma Aishwariya

2848. Fruit cracking disorder of pomegranate is seen in .... bahar

(a) Mrig bahar (b) Ambe bahar

(c) Hasth (d) None

2849. Date palm is a .... crop

(a) Monoecious **(b) Dioecious**

(c) Andromonocious (d) Gynodioecious

2850. Fruit type of date palm

(a) One seeded berry (b) Simple fruit

(c) Aggregate fruit (d) Multiple fruit

2851. Which species of *Phoenix* is found in India...?

(a) *P. indica* **(b) *P. sylvestris***

(c) *P. dactylifera* (d) All

2852. Wild species of datepalm is ....

(a) *P. indica* **(b) *P. humulis***

(c) *P. sylvestris* (d) *P. dactylifera*

2853. Metaxenia is common problem in ....

(a) Papaya (b) Datepalm

(c) a & b both (d) None

2854. .... Stage of date palm is useful for preparation of chhuhara

(a) Pind **(b) Doka**

(c) Dang (d) Tamar

2855. In India date is harvested at .... stage

(a) Dang **(b) Doka**

(c) Pind (d) None

2856. Origin of Phalsa

(a) India (b) China

(c) Africa (d) Australia

2857. Sharbati, a variety of peach is also a variety of ...

(a) *Grewia subinequalis* (b) *Carissa carandas*

(c) *Syzygium cuminii* (d) *Zizyphus mauritiana*

2858. Purple colour of jamun fruit is due to

(a) Anthocynin (b) Xanthophyll

(c) Chlorophyll (d) None

2859. Responsible for flavour of jamun fruit is due to

(a) Dihydrocarvyl acetate (b) Geranyl butyrate

(c) Terpenyle valerate **(d) All**

2860. Ber recorded which type of incompatibility

(a) Heteromorphic (b) Sporophytic

(c) Gametophytic (d) Homomorphic

2861. Which species of jamun is resistant to termite

(a) *S. densiflorum* (b) *S. jambose*

(c) *S. aqueum* (d) *S. cuminii*

2862. Jamun is commercially propagated through

(a) Seed **(b) Budding**

(c) Grafting (d) Cutting

2863. Fruit maturing time of carambola

(a) 80-90 days **(b) 100-110 days**

(c) 90-100 days (d) 110-120

2864. ……….. is generally grown in north India

(a) Ra Jamun (b) NJ- 6

(c) Paras (d) None

2865. Major problem in carambola is

(a) Heterostyly (b) Fruit cracking

(c) Stone fruit (d) None

2866. Poor fruit set is problem in annona due to …………..

(a) Cleistogamy (b) Homogamy

(c) Dichogamy (d) None

2867. Family of Bael is …..

(a) Rutaceae (b) Rosaceae

(c) Anacardiaceae (d) None

2868. Family of *Morus alba* is ....

(a) **Moraceae** (b) Myrtaceae

(c) Euphorbiaceae (d) Tiliaceae

2869. Botanical name of Tamarind

(a) *Phoenix dactylifera* (b) ***Tamarindus indica***

(c) *Morus alba* (d) *Feronia limonica*

2870. Tamarind is rich in which of the following organic acid

(a) Mallic (b) Citric

(c) **Tartaric** (d) None

2871. Tamarind fruit is botanically

(a) Pome (b) Berry

(c) **Lomentum (Pod)** (d) Drupe

2872. A following which is variety of carambola

(a) **Icambola** (b) Yogeshwar

(c) Urigam (Tamarind) (d) PKM 1

2873. *Malphigia puncifolia* belongs to family

(a) Anacardiaceae (b) **Malphigiaceae**

(c) Guttiferae (d) Anacardiaceae

2874. Most commonly used rootstock of lasoda is .

(a) ***Cordia gharaf*** (b) *Cordia mixa*

(c) Quince (d) Rose apple

2875. Chromosome number of bael

(a) 2n = 36 (b) 2n = 24

(c) 2n = 32 (d) **None**

2876. *Cordia mixa* belongs to family

(a) Tiliaceae (b) **Boraginaceae**

(c) Rutaceae (d) Euphorbiaceae

2877. Pond apple is known as...........

(a) Semi-deciduous (b) Deciduous

(c) Evergreen (d) None of above

2878. Which fruit is considered to be most delicious in annonaceous family?

(a) Ramphal (b) Lakshman phal

(c) Hanuman phal **(d) Sitaphal**

2879. Phalsa is

(a) Cleistogamous **(b) Homogamous**

(c) Dichogamous (d) Self incompatible

2880. Basic chromosome number of wood apple is..............

(a) 7 (b) 8

(c) 9 (d) 18

2881. Wood apple is generally propagated by............

(a) T-budding (b) Patch budding

(c) Seed (d) Hard wood cutting

2882. At maturity of pond apple the flesh turns in to ....

(a) Orange (b) Yellow

(c) White (d) Red

2883. Type of dichogamy present in cherimola is..............

(a) Protandry **(b) Protogyny**

(c) Hetrodichogamy (d) Hetrostyly

2884. Lakshman phal is a cross between..............

(a) *A. reticulata x A. squamosa* (b) *A. reticulata x A. cherimola*

(c) *A. squamosa x A. reticulata* (d) *A. squamosa x A. cherimola*

2885. Among following which is the variety of pomegranate ....

(a) Mridula (b) Arka Mridula

(c) Ruby **(d) (a) and (c) both**

2886. Fruit morphology of custard apple is...................

(a) **Aggregate** (b) Multiple

(c) Berry (d) Modified berry

2887. Pond apple is native of......

(a) **Tropical America - West Africa**

(b) China

(c) West America

(d) Iraq

2888. Chromosome number of phalsa is...........

(a) 18 (b) 32

(c) **36** (d) 40

2889. Edible part of wood apple is..............

(a) **Succulent placentae** (b) Pericarp

(c) Mesocarp (d) Succulent seed coat

2890. Which rootstock is used for annona?

(a) ***Annona glabra*** (b) *Annona cherymola*

(c) *Annona atemoya* (d) None of the above

2891. Which of the following is improved variety of Annona?

(a) Arka Jeet (b) **Arka Sahan**

(c) Pusa Sahan (d) Arka Navin

2892. Phalsa can be grown as an intercrop.................

(a) Bael (b) Aonla

(c) Ber (d) **All of above**

2893. Viviparous germination is found in ....

(a) **Wood apple** (b) Phalsa

(c) Carambola (d) Ber

2894. Cherimoya fruit is normally in shape...............

(a) Round (b) Irregular

(c) Heart (d) Square

2895. Self-life of custard apple is...............

(a) 4 (b) 6

(c) 8 **(d) 2**

2896. Cherimoya seed oil contains is..................

(a) Oleic acid (b) Linoleic acid

(c) Palmitic acid **(d) All of above**

2897. Which variety of aonla is suitable for candy and preserve making?

(a) NA 4 **(b) NA 9**

(c) NA 5 (d) NA 10

2898. Which variety of aonla is free from necrosis?

(a) Krishna (NA-4) (b) Kanchan (NA-5)

(c) Amrit (NA-7) **(d) Neelam**

2899. Planting time of aonla ....

(a) June – Sep. **(b) Feb.-March**

(c) Nov – Dec (d) None

2900. Pruning time of ber.....................

(a) July – Sep (b) Aug-Nov

(c) Nov-March **(d) April -May**

2901. Which variety of ber is released from CHES, Godhra...........

(a) Goma Aishwarya (b) Goma Yashi

(c) Goma Priyanka **(d) Goma Kirti**

2902. Ideal time for sowing phalsa is..............

(a) Jan –Feb (b) March – April

(c) May – June (d) July – Aug

2903. Fruits of phalsa are borne on ....

(a) 1 year old (b) 2 year old

(c) Current season growth (d) 3 year old

2904. The physiological disorder of custard apple.............

(a) Mummification (b) Fruit Cracking

(c) Crocodile skin **(d) All of the above**

2905. Washington is the variety of which crop 1. Papaya 2. Sweet Orange 3. Custard apple

(a) 1 & 2 (b) 1 & 3

(c) 2 & 3 **(d) All the above**

2906. 'Foreigner Litchi' is the synonym of which fruit crop?

(a) Guava (b) Aonla

(c) Annona (d) None

2907. Cherimoya fruit has aprophine alkaloids such as ....

(a) Romerine **(b) Annonine**

(c) Dehydromerine (d) All the above

2908. Which of the following is non climacteric fruit?

(a) Pomegranate (b) Ber

(c) Jamun **(d) All the above**

2909. The most perishable fruit crop is ....

(a) Custard apple (b) Phalsa

(c) Wood apple (d) Caramada

2910. Which fruits are cauliflоros and polyembryony?

(a) Carambola & Jamun (b) Cocoa & Ber

(c) Mango & Karonda (d) Citrus & Aonla

2911. Regular pruning is not required in which crops?

(a) Jamun (b) Aonla

(c) Sapota **(d) All the above**

2912. Which fruit contains oxalic acid?

(a) **Carambola** (b) Tamarind

(c) a) and b) (d) None

2913. Most common method of propagation in jamun?

(a) Seed (b) Inarching

(c) Grafting **(d) Budding**

2914. Origin of Jamun............

(a) Indo – Malaya. (b) Indo – Burma

(c) Indo – China (d) Indo-Japan

2915. Carambola is mostly propagated through......................

(a) Seed (b) Grafting

(c) Cutting (d) Layering

2916. Botanical name of Kair is ....

(a) *Capparis apphylla* (b) *Carissa carandas*

(c) *Morus alba* (d) *Cordia mixa*

2917. Somatic chromosome number of mulberry..................

(a) 308 (b) 90

(c) 45 (d) 56

2918. Propagation method of mulberry is ....

(a) Cutting (b) Budding

(c) Grafting (d) Seed

2919. Origin of *Cordia mixa* is ....

(a) India **(b) America**

(c) China (d) Japan

2920. Auto tetraploid variety of ber is ....

(a) Gola (b) Illaichi

(c) Umran (d) a) and b)

2921. Arka Neelachal and Vikram are the variety of custard apple released from...

(a) **Bhubaneshwar, Odisha** (b) Cettali, Karnataka

(c) Godhara, Gujarat (d) None

2922. The growth curve observed in date palm is ....

(a) **Single** (b) Double

(c) Triple (d) None

2923. Common name of Jamun is....................

(a) **Java Plum** (b) Japan plum

(c) Chinese date (d) All of the above

2924. Which Ber variety resistant to fruit fly

(a) Mehrun (b) Dodhia

(c) Sanaur- 2 (d) **a & b both**

2925. Which *carrisa spp*. suitable for jam making?

(a) *C. edulis* (b) ***C. ovata***

(c) *C. spindrum* (d) *C. bispinosa*

2926. Planting distance of tamarind

(a) 6 x 6 m (b) **12 x 12 m**

(c) 8 x 8 m (d) 1 x 1 m

2927. A trifoliate leaf is noted in...

(a) ***Aegle marmelos*** (b) *Cordia mixa*

(c) *Carissa ovate* (d) *Tamarindus indica*

2928. Which species of moraceae family cultivated for silkworm rearing?

(a) *Morus rubra* (b) ***Morus alba***

(c) *Morus nigra* (d) a & b both

2929. *Ziziphus jujube* Mill is native to...

(a) India (b) Africa

(c) **China** (d) None

2930. Which ber is thornless and evergreen...?

(a) Indian ber **(b) Chinese ber**

(c) a & b both (d) None

2931. Poona fig is planted at distance...

(a) 2.5 x 2.5 m (b) 4.5 x 4.5 m

(c) 2 x 2 m **(d) 5 x 5 m**

2932. Which crop has great tolerant to salinity, sodicity and alkalinity?

(a) Ber (b) Bael

(c) Aonla (d) Annona

2933. Maximum area under datepalm is...

(a) Banashkantha, Gujarat (b) Maheshana, Gujarat

(c) Kuchh, Gujarat (d) Disa, Gujarat

2934. Nick name of Bael

(a) Forbidden fruit

(b) Java plum, Christ-thorn

(c) Indian blackberry, Java plum

(d) Golden apple, Indian quince

2935. Self pollinated species are also called as

(a) Autogamy species (b) Autoploidy species

(c) Inbreeders (d) **Both (A) and (B)**

2936. The mechanisms, which promote self pollination is

(a) Homogamy (b) Chasmogamy

(c) Cleistogamy **(d) All of above**

2937. The species who tolerate to inbreeding is

(a) Cross pollinated species

(b) Self pollinated species

(c) Vegetatively propagated species

(d) None of the above

2938. Inbreeders are less adaptability due to

(a) Broad genetic base (b) Heterozygosity

(c) Narrow genetic base (d) None of the above

2939. Pure line is developed in

(a) Cross pollinated species

(b) Self pollinated species

(c) Vegetatively propagated species

(d) None of the above

2940. Inbred line is developed in

(a) Cross pollinated species

(b) Self pollinated species

(c) Vegetatively propagated species

(d) None of the above

2941. Clone is developed in

(a) Cross pollinated species

(b) Self pollinated species

(c) Vegetatively propagated species

(d) None of the above

2942. Cross pollinated species are also called as

(a) Allogamy species **(b) Both (A) and (B)**

(c) Outbreeders (d) Alloploidy species

2943. The mechanisms, which promote cross pollination is

(a) Hetrostyly (b) Self incompatibility

(c) Male sterility **(d) All of the above**

2944. The mating in which each male is equal chance to mate with each female is called

(a) Fmasculation **(b) Random mating**

(c) Anthesis (d) Restricted mating

2945. The removal of male part from the female parent to promote cross pollination is called

(a) Anthesis **(b) Emasculation**

(c) Pollination (d) None of above

2946. Inbreeding depression is found in

(a) Autogamous species

(b) Vegetatively propagated species

(c) Allogamous species

(d) Both (B) and (C)

2947. The first opening of the flowers is called as

(a) Anthesis (b) Emasculation

(c) Pollination (d) None of above

2948. Cross pollinated species are more adaptability due to

(a) Broad genetic base (b) Homozygosity

(c) Narrow genetic base (d) None of the above

2949. The species who propagate by other then seeds are called as

(a) Self pollinated species (b) Cross pollinated species

(c) Asexually propagated species (d) None of the above

2950. Asexually propagated species have

(a) Broad genetic base (b) More flexibility

(c) Wide adaptability **(d) All of the above**

2951. The transfer of pollen to stigma is called as

(a) Anthesis (b) Emasculation

(c) Pollination (d) None of above

2952. The F_1 crossed with one of its parent is called as

(a) Back cross (b) Top cross

(c) Test cross (d) Double top cross

2953. The F_1 crossed with a recessive parent is called as

(a) Back cross (b) Top cross

(c) Test cross (d) Double top cross

2954. The inbred line crossed with an Open pollinated variety is called as

(a) Back cross **(b) Top cross**

(c) Test cross (d) Double top cross

2955. The F_1 crossed with an open pollinated variety is called as

(a) Back cross (b) Top cross

(c) Test cross **(d) Double top cross**

2956. The full heterosis can be exploited in

(a) Hybrid (b) Synthetic variety

(c) Pureline Variety (d) Composite variety

2957. The partial heterosis can be exploited in

(a) Hybrid

(b) Synthetic & Composite variety

(c) Pureline Variety

(d) Open pollinated variety

2958. The hybrid is called as

(a) Homogeneous population (b) Heterozygous population

(c) Heterogeneous population **(d) Both (A) and (C)**

2959. The pureline variety is called as

(a) Homogeneous population

(b) Homozygous population

(c) Heterogeneous population

(d) Both (A) and (C)

2960. The first progeny of cross between the two dissimilar parents is called as

(a) Pureline **(b) Hybrid**

(c) Inbred line (d) Open pollinated variety

2961. The composite and synthetic variety population is called as

(a) Homozygous & Homogeneous population

(b) Heterozygous and Heterogeneous population

(c) Homozygous and Heterogeneous population

(d) Heterozygous and Homogeneous population

2962. The choice of breeding methods totally depends upon

(a) Mode of pollination (b) Gene action

(c) Mode of reproduction **(d) All of the above**

2963. Plant introduction is also called as

(a) General breeding methods

(b) Population improvement approaches

(c) Special breeding methods

(d) None of the above

2964. Transgenic breeding is also called as

(a) General breeding methods

(b) Population improvement approaches

(c) Special breeding methods

(d) None of the above

2965. Biparental mating is also called as

(a) General breeding methods

(b) Population improvement approaches

(c) Special breeding methods

(d) None of the above

2966. Progeny selection is used in
(a) Self pollinated crops
(b) Vegetatively propagated crops
(c) Cross pollinated crops
(d) None of the above

2967. Pureline selection is applicable only in
(a) Self pollinated crops
(b) Vegetatively propagated crops
(c) Cross pollinated crops
(d) None of the above

2968. Clonal selection is applicable only in
(a) Self pollinated crops
(b) Vegetatively propagated crops
(c) Cross pollinated crops
(d) None of the above

2969. Back cross method is used in
(a) Self pollinated crops
(b) Vegetatively propagated crops
(c) Cross pollinated crops
(d) All of the above

2970. Multiline approaches is used in
(a) Self pollinated crops
(b) Vegetatively propagated crops
(c) Cross pollinated crops
(d) None of the above

2971. Genetic variability is created through
(a) Mutation (b) Polyploidy breeding
(c) Hybridization **(d) All of the above**

2972. The gene transferring from wild to cultivated species by using

(a) Plant Introduction
(b) Domestication
(c) Wide/Distant Hybridization
(d) None of the above

2973. The concept of ideotype was first proposed by

(a) Rissell (1978)
(b) Flor (1956)
(c) Golden (1939)
(d) Donald (1968)

2974. The transposition of crop plants from the place of their cultivation to such areas where they were never grown earlier is called as

(a) Selection
(b) Domestication
(c) Plant Introduction
(d) Acclimatization

2975. The new plant material is obtained through

(a) Personal visit
(b) Correspondence
(c) Exchange of material
(d) All of the above

2976. The vegetatively propagated crops is introduced from the other countries in the form of

(a) Cutting
(b) Both (A) and (C)
(c) Propagules
(d) Seeds

2977. The important sources of plant collections is

(a) Centres of diversity
(b) Farmer's field
(c) Gene sanctuaries
(d) All of the above

2978. Primary introduction is also called as

(a) Direct introduction
(b) Secondary introduction
(c) Indirect introduction
(d) None of the above

2979. Secondary introduction is also called as

(a) Direct introduction
(b) Primary introduction
(c) Indirect introduction
(d) None of the above

2980. Introduction that can be used for commercial cultivation as a variety without any change in the original genotype is called as

(a) Direct introduction
(b) Primary introduction
(c) Indirect introduction
(d) Both (A) and (C)

2981. Any foreign variety which is directly recommended for commercial cultivation in the new environment is called as

(a) Primitive variety **(b) Exotic variety**

(c) Indigenous variety (d) Open pollinated variety

2982. Plant introduction should be used for

(a) Economic use (b) Study of origin

(c) Conservation of diversity **(d) All of the above**

2983. In India, plant introduction for different crop plants is carried out by

(a) IPGRI

(b) Biodiversity Internationals

(c) NBPGR

(d) None of the above

2984. The head quarter of Biodiversity international is

(a) Rome (b) USA

(c) India (d) UK

2985. The agencies which are responsible for plant introduction in India are,

(a) NBPGR (b) BSI

(c) FRI **(d) All of above**

2986. The new name of IPGRI is

(a) Biodiversity international (b) Biodiversity

(c) Diversity international (d) Biosafty international

2987. The agency deals with introduction of forest tree in India is,

(a) NBPGR (b) BSI

(c) FRI (d) BI

2988. The agency deals with the introductions of medicinal and botanical plants in India is

(a) NBPGR **(b) BSI**

(c) FRI (d) BI

2989. The adaptation of introduced variety to the new environment is called as

(a) Selection (b) Domestication

(c) Introduction **(d) Acclimatization**

2990. The process of bringing wild and weedy species under human management is called as

(a) Selection **(b) Domestication**

(c) Introduction (d) Acclimatization

2991. The introduced variety is used as

(a) A variety (b) A variety after selection

(c) A parent in hybridization **(d) All of the above**

2992. The term heterosis was coined by

(a) East (1908) (b) Jones (1917)

(c) Shull (1914) (d) Devenport (1908)

2993. The terms horizontal and vertical resistance were first used by

(a) Jennings **(b) Van Der Plant (1963)**

(c) Thoday (d) Robinson (1971)

2994. The development of seed without sexual process is

(a) Polyembryony **(b) Apomixis**

(c) Metaxenia (d) Xenia

2995. The progeny test was developed by

(a) Vilmorin (b) Le Couteur

(c) Johannsen (d) Hallet

2996. The concept of pureline was proposed in 1903 by

(a) Vilmorin (b) Le Couteur

(c) Johannsen (d) Hallet

2997. Self pollination increase

(a) Homozygosity (b) Homogeneity

(c) Heterozygosity (d) Heterogeneity

2998. Purity of existing pureline varieties can be maintained by regular

(a) Pureline selection **(b) Mass selection**

(c) Pedigree selection (d) Recurrent selection

2999. Purelines can be used as/for

(a) As varieties (b) For studies in mutation

(c) As parents in hybridization **(d) All of the above**

3000. Pureline selection has been the most commonly used method in the early stage for improvement of

(a) Allogamous crops

(b) Often cross pollinated crops

(c) Autogamous crops

(d) Both (B) and (C)

3001. Pureline selection can be applied for the improvement of

(a) Local varieties (b) Old pureline varieties

(c) Introduced varieties **(d) All of the above**

3002. The methods used for handing of segregating generations are

(a) Bulk method (b) Pedigree method

(c) Back cross method **(d) All of the above**

3003. The modified form of a bulk method is

(a) Pedigree method

(b) Single Seed Descent method

(c) Back cross method

(d) Mass selection

3004. The method which can be used for transferring oligogenic character without altering the original genotypes

(a) Bulk method (b) Pedigree method

(c) Backcross method (d) SSD method

3005. The lines with a single gene differences is called as

(a) Pure lines (b) RILs

(c) Isogenic lines (d) None of above

3006. Isogenic line can be produced by using

(a) Bulk method (b) Pedigree method

(c) Backcross method (d) SSD method

3007. The term "Boom and Bust Cycle" was suggested by

(a) Marshall **(b) Priestley**

(c) Browning (d) BorLaug

3008. The frequency of spontaneous mutation is

(a) 10^{-4} **(b) 10^{-6}**

(c) 10^{-5} (d) 10^{-7}

3009. The oldest method of selection is

(a) Recurrent selection (b) Ear-to-row selection

(c) Mass selection (d) Mass-pedigree selection

3010. Mass selection is based on

(a) Genotypes only (b) Progeny test only

(c) Phenotypes only (d) Test cross only

3011. Selection is ineffective

(a) Within purelines

(b) Within Composite varieties

(c) Between purelines

(d) Between composite varieties

3012. Johannsen (1903) developed pureline concept using

(a) Garden pea (b) Indian bean

(c) Common bean (d) Onion

3013. Mass selection is applicable to

(a) Cross pollinated species

(b) Self pollinated species

(c) Vegetatively propagated species

(d) Both (A) and (C)

3014. In self pollinated crops, the constitution of mass selected variety is

(a) Homozygous & Homogeneous population

(b) Heterozygous and Heterogeneous population

(c) Homozygous and Heterogeneous population

(d) Heterozygous and Homogeneous population

3015. When desirable plants are selected from a mixed population and their seeds are mixed to gather to grow next generation is called

(a) Positive mass selection (b) Pureline selection

(c) Negetive mass selection (d) Pedigree selection

3016. Mass selection is more effective when the heritability is

(a) Moderate **(b) High**

(c) Low (d) None of above

3017. The grid method of mass selection is proposed by

(a) East (1908) (b) Jones (1917)

(c) Gardener (1962) (d) Devenport (1908)

3018. Mass selection can take atleast ________ years for releasing a variety

(a) 11-12 years **(b) 6-7 years**

(c) 9-10 years (d) 14-15 years

3019. Progeny test should be done in

(a) Pureline selection **(b) Progeny selection**

(c) Mass selection (d) Back cross method

3020. Progeny of a single plant obtained by asexual reproduction is called as

(a) Pureline **(b) Clone**

(c) Inbred line (d) None of above

3021. Clones are naturally.

(a) Homozygous (b) Hemizygous

(c) Heterozygous (d) None of above

3022. Segregation was not found in

(a) Purelines (b) Synthetic varieties

(c) Composite varieties (d) None of the above

3023. The clone should be deteriorate by

(a) Mutation (b) Viral diseases

(c) Bacterial disease **(d) All of the above**

3024. Pedigree method is used to improve

(a) Oligogenic trait (b) Both (A) and (B)

(c) Polygenic trait (d) None of the above

3025. Selection is applied in bulk method from

(a) F_2 onward (b) F_5 onward

(c) F_3 onward **(d) F_6 onward**

3026. The chance of obtaining transgressive segregation is more in

(a) Plant Introduction (b) Polyploidy breeding

(c) Mutation breeding **(d) Bulk breeding method**

3027. The number of backcrosses are sufficient to retain the genotypes of original variety with new characters

(a) 3 to 4 backcrosses (b) 4 to 5 backcrosses

(c) 5 to 6 backcrosses (d) 10 to 11 backcrosses

3028. The first artificial hybrid was proposed by

(a) Joseph Koelreuter (b) Thomas Andrew Knight

(c) Thomas Fairchild (d) Mendel

3029. The first artificial plant hybrid was produced through a cross between

(a) Two varieties of rice

(b) Two varieties of pea

(c) Two varieties of maize

(d) Sweet Willam and Carnation

3030. Intervarietal hybridization is also called as

(a) Interspecific hybridization (b) Distant hybridization

(c) Intraspecific hybridization (d) Intergeneric hybridization

3031. Thomas Andrew knight used hybridization to develop popular varieties of

(a) Apple (b) Grape

(c) Peach **(d) All of above**

3032. Selection for simply inherited character is started from

(a) F_2 (b) F_4

(c) F_3 (d) F_5

3033. Which emasculation method is suitable for accurate genetic studies?

(a) Self incompatibility **(b) Hand emasculation**

(c) Male sterility (d) Hot water treatment

3034. The information is recorded on tags during hybridization is

(a) Date of emasculation (b) Name of the parents

(c) Date of pollination **(d) All of the above**

3035. An Intergeneric hybrid Raphanobrassica was produced by

(a) Karpechenko (b) Thomas Fairchild

(c) Lysenko (d) Koelreuter

3036. The first man made cereal "Triticle" was developed by

(a) Karpechenko (b) Thomas Fairchild

(c) Rimpau (d) Koelreuter

3037. Chromosome number of any plant species can be doubled by the use of

(a) Tissue culture (b) Nitrous oxide

(c) Colchicine **(d) All of above**

3038. The full form of "EBN" is

(a) Economic botany of Nicotiana

(b) Experimental biology of Nicotiana

(c) Endosperm balance number

(d) Embryo balance number

3039. The most common agent used for chromosome doubling is

(a) Nacl (b) Callus

(c) Nitrous oxide **(d) Colchicine**

3040. The important and most critical factor for hybridization is

(a) Choice of parents (b) Pollination

(c) Emasculation (d) Anthesis

3041. In hybridization programme, the most important consideration for choice of parents is

(a) Per se performance (b) Local adaptation

(c) Combining ability (d) Genetic diversity

3042. The prime purpose of emasculation is

(a) Promote crossing (b) Promote seed act

(c) Prevent selfing (d) Prevent out crossing

3043. The most widely used emasculation method is

(a) Cold treatment (b) **Hand emasculation**

(c) Alcohol treatment (d) Hot water treatment

3044. Sterility is developed in wide hybridization may be due to

(a) Lack of chromosome pairing (b) Cytoplasm of alien species

(c) Cryptic structural hybridity **(d) All of the above**

3045. The breeding method is most suited to handle distant crosses

(a) DSM (b) Pedigree

(c) SSD **(d) Backcross method**

3046. The first step in genetic improvement of a plant species is

(a) Domestication (b) Germplasm collection

(c) Selection (d) Hybridization

3047. Which is a prerequisite for crop improvement?

(a) Easy pollination control (b) Cytoplasmic male sterility

(c) Genetic variation (d) Sexual reproduction

3048. The cocept of centers of origin of crop plants was proposed by

(a) J.R. Harlan (b) C. Linnaeus

(c) N.I Vavilov (d) J.B. Hutchinson

3049. A crop showing less than 5% cross pollination is considered as

(a) Self-pollinated (b) Cross-pollinated

(c) Often-cross pollinated (d) None of the above

3050. A crop showing 5 to 30% cross pollination is considered as

(a) Self-pollinated (b) Cross-pollinated

(c) Often-cross pollinated (d) None of the above

3051. Nullisomics are represented by which of the following?

(a) 2n - 1 – 1 (b) 2n + 1 – 1

(c) 2n – 2 (d) 2n– 2 +1

3052. The basic chromosome is denoted by

(a) n (b) 3n

(c) 2n **(d) x**

3053. The haploid chromosome is denoted by

(a) n (b) 3n

(c) 2n (d) x

3054. The diploid chromosome is denoted by

(a) n (b) 3n

(c) 2n (d) x

3055. In true diploid species

(a) n = x (b) n < x

(c) n > x (d) n => x

3056. The condition in which the numerical changes in entire genome is called as

(a) Heteroploidy (b) Aneuploidy

(c) Euploidy (d) None of above

3057. Any change in the chromosome number from the diploid state is called as

(a) Heteroploidy (b) Aneuploidy

(c) Euploidy (d) None of above

3058. The change in one or few chromosome of the genome is called as

(a) Heteroploidy **(b) Aneuploidy**

(c) Euploidy (d) None of above

3059. The development of embryo from egg cell without fertilization is called as

(a) Emasculation (b) Apospory

(c) Apogamy **(d) Parthenogenesis**

3060. Development of more than one embryo in seed is known as

(a) Monoembryony (b) Both (A) and (B)

(c) Polyembryony (d) None of the above

3061. Monosomic is represented by

(a) 2n – 1 (b) 2n + 1 – 1

(c) 2n – 2 (d) 2n– 2 +1

3062. Trisomic is represented by

(a) 2n – 1 **(b) 2n + 1**

(c) 2n – 2 (d) 2n– 2 +1

3063. Tetrasomic is represented by

(a) 2n – 1 (b) 2n + 1

(c) 2n – 2 **(d) 2n+2**

3064. Polyploids which originate by multiplication of the chromosome of a single species are known as

(a) Autopolyploidy (b) Heteroplidy

(c) Allopolyploidy (d) Both (A) and (B)

3065. Autotriploids can be developed through

(a) Haploid X Haploid (b) Diploid X Diploid

(c) Haploid X Diploid **(d) Diploid X Tetraploid**

3066. Apple is

(a) Allotriploid (b) Autotetraploid

(c) Autotriploid (d) Autohexaploid

3067. Graps is

(a) Allotriploid **(b) Autotetraploid**

(c) Autotriploid (d) Autohexaploid

3068. Allopolyploid species is also known as

(a) Hybrid polyploids (b) Multispecies polyploids

(c) Bispecies polyploids **(d) All of the above**

3069. Incorporation of two or more genes in the host for specific resistance of pathogen is called as

(a) Gene pyramiding (b) Gene mining

(c) Gene deployment (d) Gene farming

3070. The gene pyramiding approach has been used first in Oat against

(a) Black rust **(b) Crown rust**

(c) Brown rust (d) Powdery mildew

3071. The s trategic use of major genes in the development of resistant cultivars for various geographical areas called as

(a) Gene pyramiding (b) Gene mining

(c) Gene deployment (d) Gene farming

3072. In Insect resistance, which method is not true resistance mechanism?

(a) Non preference (b) Tolerance

(c) Antibiosis **(d) Avoidance or Escape**

3073. The resistance mechanism non preference is also called as

(a) Antixenosis (b) Antibiosis

(c) Antioxident (d) None of above

3074. Various features of host plant that make the host unattractive to insect for food or reproduction is called as

(a) Non preference (b) Tolerance

(c) Antibiosis (d) Avoidance or Escape

3075. The adverse effect of host plant on the development and reproduction of insect pests which feed on resistant plant is called as

(a) Non preference (b) Tolerance

(c) Antibiosis (d) Avoidance or Escape

3076. The ability of a variety to produce greater yield than susceptible variety at the same level of insect attack is called as

(a) Non preference **(b) Tolerance**

(c) Antibiosis (d) Avoidance or Escape

3077. For insect resistance, what are the different sources?

(a) Cultivated varieties (b) Wild species

(c) Mutations **(d) All of the above**

3078. In insect, new biotypes have been developed through

(a) Spontaneous mutation **(b) Both (A) and (B)**

(c) Hybridization (d) None of the above

3079. Typical example of parthenocarpy

(a) Mango (b) Banana

(c) Mangosteen **(d) b & c**

3080. Monogenic resistance is __________ than oligogenic and polygenic resistance.

(a) Lesser durable (b) Moderately durable

(c) Highly durable (d) Both (B) and (C)

3081. In crystallized fruit, the spoilage is mainly due to

(a) Fermentation (b) Oxidation

(c) Insects (d) Air

3082. Fruits like amla and myrobalan are kept in a strong solution of common salt to remove

(a) Sugars **(b) Astringency**

(c) Toxic compounds (d) Acidity

3083. The ratio of the cane sugar and water in making syrup for use in glazing is

(a) 2:1 (b) 1:2

(c) 3:1 (d) 1:3

3084. The equivalent weight of citric acid is

(a) 60.04 (b) 74.04

(c) 70.04 **(d) 64.04**

3085. Low acid foods have a pH

(a) < 5.0 **(b) >5.0**

(c) <5.5 (d) >5.5

3086. What is used for the determination of jelly grade by measuring relative viscosity of pectin

(a) Gelmeter (b) Jelmeter

(c) Viscometer (d) Pycnometer

3087. Grape juice, apple juice and lime juice cordial should have

(a) Brilliant clear appearance (b) Dull appearance

(c) Sparkling appearance **(d) Both a & c**

3088. Lime juice is kept for 3-6 month in carboy or barrel with addition of 700 ppm SO_2 for

(a) Filtration **(b) Clarification by setting**

(c) Fermentation (d) None

3089. Freezing is used for clarification of which of the following pasteurized juice

(a) Grapes juice (b) Mango

(c) Passion fruit (d) Grape fruit juice

3090. Which of the following enzymes is used for clarification of the juice

(a) Pectinol (b) Papain

(c) Amylase (d) Clarizol

3091. Clarification of sugar added is done by adding small quantity of

(a) citric acid (b) few drop of lime juice

(c) Ascorbic acid **(d) both a & b**

3092. Which acid is often used in fruit drinks

(a) Citric acid (b) Phosphoric acid

(c) Tartaric acid (d) ascorbic acid

3093. Replacement of the nutrient lost during processing is known as

(a) Fortification (b) Scalding

(c) Enrobing **(d) both a & c**

3094. Which of the juice can be clarified by low temperature storage for months

(a) Apple **(b) Grape**

(c) Litchi (d) Plum

3095. Pomegranate juice can be preserved by addition

(a) **Sodium benzoate** (b) Sulphur dioxide

(c) Paraben (d) KMS Gelatin

3096. Gelatin can be used as a

(a) Sequestering agent (b) **Clarifying agent**

(c) Nutrient supplement (d) None

3097. RTS should contain fruit juice, %TSS and

(a) 0.5% acid (b) **0.3% acid**

(c) 1% acid (d) 1.2% acid

3098. Nectar should contain 20% fruit juice and

(a) 15% TSS and 0.5% acid (b) 12% TSS and 0.5% acid

(c) **15% TSS ami 0.3% acid** (d) 15% TSS and 1 % acid

3099. Which of the following beverage is suitable for blending with wine

(a) **Cordial** (b) Syrup

(c) Squash (d) Apptizer

□□□

2

Fill in the Blanks

1. ___________ is a climacteric fruit.

   (**Mango**, Grape, Citrus, Pomegranate)

2. ___________ is a non-climacteric fruit.

   (Banana, **Pomegranate**, Sapota, Guava)

3. ___________ is long day plant.

   (Strawberry, Pineapple, Coffee, **Apple**)

4. ___________ is a short day plant.

   (Passion fruit, Papaya, Apple, **Strawberry**)

5. ___________ is a day neutral plant.

   (Strawberry, Banana, **Guava**, Pineapple)

6. ___________ is a evergreen plant.

   (**Sapota**, Grapes, Guava, Fig)

7. ___________ is a deciduous plant.

   (**Ber**, Mango, Banana, Areca nut)

8. Deficiency of Vitamin A causes ___________.

   (**Night blindness**, Beriberi, Scurvy, Common cold)

9. ___________ system accommodates 16 % more tree per hectare as compared to square system.

   (**Hexagonal**, Triangular, Rectangular, Diagonal)

Bold Means Correct Answer

10. __________ system is followed in hilly or undulated area.

(**Contour**, Square, Triangular, Hexagonal)

11. __________ is the practice of covering soil around plants.

(**Mulching**, Fumigation, Solarization)

12. __________ is a practice of growing two or more crops on the same piece of land in one calendar year.

(**Multiple cropping**, Mono cropping, Inter cropping, Mixed cropping)

13. __________ define as growing of two or more crops in a particular sequence on the same piece of land with for higher production without depleting the soil fertility.

(**Crop rotation**, Ratoon cropping, Relay cropping, Mono cropping)

14. __________ is a plant growing where it is not wanted.

(**Weed**, Seed, Tree, Shrub)

15. __________ is a non selective herbicide.

(**Glyphosate**, Dalapon, 2-4-D, Fluchloralin)

16. __________ is a practice of ploughing or turning undecomposed green plant material into the soil for improving the physical conditions.

(**Green manuring**, Vermicomposting, Vermi wash, Manuring)

17. In __________ system branches are allowed to grow indefinitely, so that it will grow more rapidly and vigorously than the side branches.

(**Central leader**, Open centre, Modified leader, Bush system)

18. __________ plant growth regulator is used as weedicide/Herbicide.

(**Auxin**, Gibberellin, Cytokinin, Ethyelene)

19. In __________ irrigation method water is directly applied to the root zone.

(**Drip**, Sprinkler, Border, Check Basin)

20. In ________irrigation method water is sprayed into air through nozzles.

    (Drip, **Sprinkler**, Flood, Border)

21. Fruits containing few or several seeds are called________.

    (**Berry**, Drupe, Pome, Capsule)

22. Highest ethylene is produced in ________fruit crop.

    (**Apple**, Mango, Banana, Sapota)

23. __________ is a highly salt tolerant crop.

    (**Datepalm**, Mango, Aonla, Citrus)

24. Richest source of Vitamin C is in __________

    (Citrus, **Aonla**, Banana, Guava)

25. Heavy pruning is done in __________ fruit crops.

    (Grapes, Ber, Fig, **All the above**)

26. Very low temperature causes __________ injury in plants.

    (**Winter**, Summer High temperature, All the above)

27. Bower/Pandal system is basically used in __________ fruit crop.

    (**Grapes**, Mango, Sapota, Guava)

28. Cool season fruit crops require __________ temperature to survive.

    (15to 20^0 C, 20 to 25^0 C, 30 to 35^0 C, **18 to 24^0 C**)

29. __________ fruit crop is polyembyonic type in nature.

    (Mango, Jamun, Citrus, **All the above**)

30. Plants which withstand minimmm temperature of -4 to -20 C are called __________ plants.

    (**Hardy**, Half Hardy, Tender, None of these)

31. __________ is a raw material for photosynthesis of plants.

    (**Water**, Sunlight, Chlorophyll, None of the above)

32. __________ is a ripening hormone.

(**Ethylene**, Giberellin, Cytokinin, All the above)

33. __________ is a stress hormone.

(**Abscissic acid**, Auxin, Cytokinin, None of the above)

34. __________ PGR is used to increase the berry size in grapes.

(**Giberrelic acid**, NAA, 2,4-D. Ehtylene)

35. Vitamin- C is commonly known as __________

(Citric acid, Acetic acid, Lactic acid, **Ascorbic acid**)

36. Which of the following is poor mans fruit.

(**Ber**, Mango, Grapes, Apple)

37. __________ is a type of micro propagation.

(Callus culture, Tissue culture, Meristem culture, **All of above**)

38. In Gujarat there are __________ types of agro-climatic zones.

(7, **8**, 9, 10)

39. __________training method followed in Mango.

(**Central leader**, Open centre, Modified central leader, Multiple system)

40. Pruning is a __________ process.

(Talling, **Dwarfing**, Shortening, All of the above)

41. __________training method followed in Sapota.

(**Central leader**, Open centre, modified central leader, Multiple system)

42. __________ is a temperate fruit crop.

(**Apple**, Mango, Sapota, Banana)

43. __________ is a tropical fruit crop.

(Apple, Almond, Pomegranate, **Mango**)

44. _________ is a sub-tropical fruit crop.

(Mango, **Pomegranate**, Coconut, Jack fruit)

45. Plants which can withstand minimum temperature of -4 to -20 C are called _________ plants.

(Hardy, **Half Hardy**, Tender, None of these)

46. _____ plants are those which can not withstand temperature of 0^0 C.

(**Tender**, Hardy, Half hardy, All the above)

47. Warm season fruit crops require _________ temperature to survive.

(**25to 35^0 C**, 20 to 25^0 C, 30 to 35^0 C, 18 to 24^0 C)

48. _________ plant is grown around the orchard to stop wind damage as a wind break.

(**Eucalyptus**, Bougainvelia, Thevetia, None of the above)

49. _________ plant is grown around the orchard as a good fence.

(Eucalyptus, **Bougainvillea**, Neem, Casurina)

50. Fruits attracting birds and animals should be near _________.

(**Watchman's hurt**, Office Building, Near road, None of the above)

51. Evergreen trees should be planted to the _________ of the orchard.

(Backside, **Front Side**, Watchman's hurt, All the above)

52. Deciduous trees should be planted to the _________ of the orchard.

(**Backside**, Front Side, Watchman's hurt, All the above)

53. In _________ planting system plant to plant and row to row distance is equal.

(Rectangular, Diagonal, **Square**, Contour)

54. In _________ planting system row to row distance is greater than plant to plant.

(**Rectangular**, Diagonal, Square, Contour)

55. The recommended planting spacing for Mango and Sapota is ________.

(**10x10m**, 10X5m, 5x5m, 2x2m)

56. The recommended spacing for Banana is ________

(**1.8x1.8m**, 2x2m, 3x3m, 4x4m)

57. Blue green algae fix significant quantities of ________ in soil.

(Phosphorus, **Nitrogen**, Potash, None of the above)

58. The thermal death point of most plant cell is around ________

(30^0 C, 40^0 C, **60^0 C**, 70^0 C)

59. ________ is a branch which deals with the scientific study of fruits.

(Olericulture, Floriculture, Tissue culture, **Pomology**)

60. ______ is a branch which deals with the scientific study of vegetables.

(**Olericulture**, Floriculture, Tissue culture, Pomology)

61. Horticultural crops are rich source of vitamins and minerals so they are called ________

(**Protective food**, Staple food, Sustainable food, None)

62. Per day consumption of fruits per adult should be ________.

(46g, **85g**, 90g, 100g)

63. Banana is a ________ type of fruit.

(**Berry**, Pome, Drupe, Nut)

64. Seed of pomegranate is called ________

(Nut, **Aril**, Drupe, Stone)

65. Cashew nut is a ________ type of fruit.

(Berry, Pome, Drupe, **Nut**)

66. Aonla is a ________ type of fruit.

(Drupe, Pome, **Capsule**, Berry)

67. Mango is a ________ type of fruit.

(**Drupe**, Pome, Capsule, Berry)

68. ________ is necessary to build bines and teeths.

(**Calcium**, Fibers, Iron, Carbohydrates)

69. ________ fruit crop increase the heamoglobin in human being.

(**Dates**, Mango, Citrus, Grapes)

70. ________ is the richest source of Vitamin A

(**Mango**, Banana, Citrus, Date palm)

71. Which of the following fruit(s) is/are monocotyledonous?

(Banana, Pineapple, Date palm, **All of the above**)

72. Botanically, Sapota fruit is known as ________.

(Pome, Drupe, **Berry**, Nut)

73. Parthenocarpic fruits are ________

(False fruit, Composite fruit, Multiple fruits, **Seedless fruits**)

74. ________ is done to remove field heat from the fruits after harvesting.

(**Pre-cooling**, Heating, Both, None of these)

75. The word horticulture is derived from 'Hortus' and 'Coleure' meaning Garden and Cultivation. These words are ________.

(**Latin**, Greek, Persian, English)

76. Hespiridium is botanical form of fruit in ________.

(**Citrus**, Apple, Mango, Coconut)

77. Best soil for planting a fruit orchard should have pH between ________

(3.5-5.5, **6.5-7.5**, 7.5-10.5, 4.5-6.5)

78. *Mangifera indica* is the botanical name of ________

(Sapota, **Mango**, Banana, Cashew nut)

79. __________ is a cross pollinated crop.

(**Mango**, Tomato, Phalsa, All)

80. __________ is a Self pollinated crop.

(**Sapota**, Date palm, Citrus, All)

81. __________ one is botanically not a fruit.

(**Cashew apple**, W. apple, Custard apple, Pine apple)

82. Aonla and Guava are the richest source of Vitamin __________.

(C, A, D, E)

83. Mango is the richest source of Vitamin __________.

(C, **A**, D, E)

84. Scurvy and common cold is caused due to deficiency of Vitamin ______.

(C, A, D, E)

85. __________ is also called as a friend of farmers.

(**Earthworm**, Silkworm, Bacteria, Rat)

86. __________ is done to give size and shape to plants.

(Pruning, **Training**, Thinning, None of the above)

87. There are __________ elements have been considered as essential for higher green plant.

(**16**, 17, 15, 14)

88. __________ is a nitrogen fixing bacteria.

(**Azotobacter**, Phosphobacter, Bacillus, Aspergillus)

89. High value Vegetables and flowers can be grown in off season in __

(**Green house**, Lath house, Open Field, None of above)

90. __________ is a highly acid tolerant crop.

(**Bael**, Pomegranate, Jamun, Phalsa)

91. __________ is a highly salt tolerant crop.
(**Date palm**, Fig, Bael, Plum)

92. Botanically Pomegranate is ________ type of fruit.
(**Balausta**, Pepo, Drupe, Berry)

93. Botanically banana is a ________ type of fruit.
(Balausta, Pepo, Drupe, **Berry)**

94. Alternate bearing habit is commonly seen in ________ fruit
(**Mango**, Pomegranate, Jamun, Phalsa)

95. Ricketts in children is caused due to deficiency of Vitamin ________
(C, **D**, E, none of the above)

96. The best breeding method adopted to disease resistance is ________
(Test cross, **Back cross**, selection, Anther culture)

97. The higest chilling requirement among temprate fruit is an ________
(**Cherry**, Almond, Apricot, Apple)

98. The acid content during ripening in banana ________
(Medium, **Increases**, Lower, None)

99. Tangelo is cross between Mandarins & ________
(S. Orange, Acid lime, **Grape fruit**, Mandarin)

100. The temprate fruit most affected by cracking is ________
(Apple, **Cherry**, Strawberry, Pear)

101. The upper salt limit of grape is ________
(**0.3%**, 0.4%, 0.5%, 0.1%)

102. The upper salt limit of citrus is ________
(0.3%, 0.4%, 0.5%, **0.1%**)

103. Viviparous seed are reported in ________
(Mango, Sapota, **Jackfruit**, Ber)

104. Pusa Nanha is __________ type of papaya.

(**Dioecious**, gynodioecious, androdioecious, monoecious)

105. Ploidy of cultivated strawberry is __________

(Polyploidy, **octaploid**, diploid, none)

106. Pond apple is a botanically **Annona glabra**

107. Pomegranate is preferably trained on **multiple stem.**

108. Prickly pear refers to fruit of **Opuntia.**

109. San Jose Scale got entry into India in Kashmir from **France in 1906**

□□□

3

Memory Based NET/ARS Questions 2013-2014

110. Chemical used for bud breaking in Grape - **Dormex**
111. PGR used for berry elongation in Grape - **GA_3**
112. Greening and granulation is associated with - **Citrus**
113. For banana important nutrient is - **K**
114. Haploid chromosome No. of Apple - **17**
115. King of Arid fruit - **Ber**
116. Origin of Jackfruit - **India**
117. Genome size of Papaya - **742.3 Mbp**
118. Genome size of Banana - **523 Mbp**
119. Dehorning technique is used in – **Mango**
120. Precursor of auxin – **Tryptophan**
121. Storage temperature for Banana- **13^{0}C**
122. Temporary method of preservation- (Asepsis/**pasteurization**/sterilization/fermentation)
123. Microorganism used in cider preparation- **yeast**/ mould/ bacteria
124. In which fruit crop seedling is dioecious- (Avocado/Mango/**Rambutan**/Litchi)
125. Propagation method of pomegranate- **HWC**
126. Training system adopted in Pomegranate- **Multi-stem**
127. Optimum temperature for development of pomegranate fruit- 26-30/20-25/12-10/**35^{0}C**

Bold Means Correct Answer

128. Which is a subtropical fruit- Mango/ Jackfruit/ **Ber** /Banana
129. Pruning is heavily required in which fruit crop- Apple/**Peach**/ Pear/ Plum
130. Training and pruning are not much require in which fruit crop- Apple/ Ber/ Pomegranate/**Papaya**
131. Summer pruning is adopted in – **Ber**
132. Molecular markers (AFLP/RFLP/RAPD) is used for – **Phylogenetic study**/ to find out unknown species/ varietal evaluation/all
133. Kiwifruit follows – single/ double/ **triple**/ none
134. Which planting system is generally adopted in Strawberry- **matted row**/ hill row/ double hill row/ undulated ridges
135. How many species are given by Tanaka (1954) in mandarin group- **35**
136. ACC synthase causes- **delay ripening**/ reduce ethylene biosynthesis/ promote ripening
137. Which is dwarfing rootstock of apple- M-7/**M-9**/ M25/ MM111
138. Which is not a character of apple- **Polyemryony**/Deciduos/ Alternate bearer/
139. Control of calyx end rot by spraying- **Ca++**
140. What should be used in case of mango malformation- **NAA 200ppm**
141. Which auxin is used as weedicide widely used- **2,4-D**/IBA/IAA/ NAA
142. Hormone also used as a weedicides- **2,4-D**
143. Which factor is not responsible for dwarfing in mango- **Weed**
144. Which parent is used in back cross breeding method – **Recurrent**
145. Tissue culture mostly adopted in which fruit crop- **Banana**
146. Undifferentiated mass of cell is called- **Callus**
147. In which protected structure growth studies are measure easily- **Phytotron**
148. Mostly used structure in India- **Green house**/ Net house/ Sheded net house/ Glass house
149. Purpose of mist chamber in propagation- grafting/budding/ **rooting of cuttings**/ layering

150. Most commonly used irrigation system in protected cultivation- **Drip**/Sprinkler

151. Highest chilling requirement in – **Cherry**/ Apple/ Peach/ Pear

152. Chilling requirement of Bartlett pear- 150/750/**1500**/250

153. Optimum thickness of LDPE for protected cultivation –**200**

154. Apple training system-

155. Stenospermocarpy found in – Apple/ **Plum** /Pomegranate/Banana

156. The most expansive protected structure – **Phytothron**

157. In which crop area is highest in India but production is second- **mango**/ banana

158. Parent of grape variety Manjari Navin- **Centenelian**

159. Marsh and foster are the varieties of – **Grapefruit**

160. In citrus classification scientist also known as Lumper- **Swingle/ Tanaka**

161. Ornithophilous mode of pollination in – Banana/ **Pineapple**

162. Variety yet not have been developed in – Litchi/ Mango/ Atemoya /**Mangosteen.**

163. Protogyny found in – Muricata/ **Atemoya.**

164. Protandry found in – **Muricata**/ Atemoya.

165. PPVFR act in- **2001** /2003 /2005 /2007.

166. Mango variety for HDP- **Amrapali**

167. Which is non climacteric fruit- **litchi**

168. Which is non climacteric fruit- Sweet **Orange**

169. What should be use in case of fruit necrosis- **0.2 % Borax.**

170. Substitute for low temperature- **GA_3**

171. Hormone involved in germination of seed- **GA_3**/Cytokinin/ ABA/ BA

172. Which hormone is high at time of water stress- GA/**ABA**/IBA/ Cytokinin

173. Anti-gibberellin in mango- **Cultar**/ 2,4D/ABA

174. Flowering induced in Mango by using- **Cultar**

175. *Passiflora edulis* is botanical name of – **Passoin fruit**

176. __________ of pineapple is highly suited for HDP- **Kew**

177. Papaya is reachest enzyme source of __________ **Papain**

178. Pectin rich fruits includes **Quience** and **Guava**

179. Mottling in citrus is due to **Zinc** deficiency

180. Pear is traine to __________ system of training –**Central Leader**

181. The HQ of International Mango Registration Society is at **IARI, N. Delhi**

182. The HQ of ISHS is at **Belgium**

183. Chromosome number of Umran cultivar of ber **96**

184. Mango flowers during **Jan-March**

185. Mangla is an improved variety of **Arecanut**

186. Mango shows __________ type of self incompatibility.- **Sporophytic**

187. Mangosteen commercially propogated by **Seed**

188. Highest production of grape is in the state in India—**Maharashtra**

189. __________ is a practice adopted in phalsa cultivation—**Pruning**

190. Phalsa is **self pollinated** crop

191. __________ is a limiting factor in papaya cultivation— — **Frost**

192. ________ is nocturnal and serious pest of citrus—**Fruit sucking moth**

193. Citron is botanically name as—***C. medica***

194. __________ citrange has aroma similar to mango –**Merton**

195. CNSL stands for -**Cashewnut Shell Liquid**

196. Apple exhibit __________ type of apomixes -**Recurrent**

197. At least ____ pollinizer are recommended in apple orchards –**33%**

198. Apomictic seedlings are also known as **Nucellar seedlings**

199. Barbados cherry is a belongs to family – **Malphigiaceae**

200. Barbados cherry, the richest source of vit –C is about **1400 mg/100g**

❑❑❑

4

Some Sample Questions

Fruit Science Questions

1. Which fruit is rich in citric acid?
   (a) Apple (b) Peach
   (c) Pear (d) Plum

2. Pineapple is which kind of fruit?
   (a) Sorosis (b) Capsule
   (c) Synconus (d) Drupe

3. Which one is not a berry fruit?
   (a) Banana (b) Papaya
   (c) Sapota **(d) Bael**

4. Which one is climateric fruit?
   (a) Pineapple (b) Ber
   (c) Annona (d) Litchi

5. Which fruit shows duodichogamy?
   (a) Avocado (PDSD) **(b) Chestnut**
   (c) Pecan nut (d) Cashew nut

6. Kiwi fruit belongs to which category?
   (a) Auto-octaploid (b) Allo-octaploid
   (c) Auto-hexaploid (d) Allo-hexaploid

Bold Means Correct Answer

7. Which citrus fruit is from the new world?

(a) Grape Fruit (b) Pummelo

(c) Lime (d) Sweet orange

8. Which fruit is called commonly "Bathroom Fruit"

(a) Banana (b) Kiwi

(c) Guava **(d) Mango**

9. Which fruit is called "Chinese Gooseberry"

(a) Fig **(b) Kiwi**

(c) Aonla (d) Strawberry

10. Which fruit is called "Indian Gooseberry"

(a) Fig (b) Kiwi

(c) Aonla (d) Strawberry

11. Which sub-tropical fruit was introduced in India under name of 'Japanese Medlar'?

(a) Peach (b) Plum

(c) Loquat (d) Quince

12. Propagation method in cashew nut

(a) Softwood grafting (b) Hardwood cutting

(c) Seedling (d) Patch budding

13. Inflorescence of arecanut is:

(a) Spadix (b) Corymb

(c) Solitary (d) Panicle

14. Edible part of banana is:

(a) Mesocarp (b) Endocarp

(c) Pericarp **(d) Both a & b**

15. Variety as a pollinizer used in mango is:

(a) Chausa (b) Mallika

(c) Bombay Green (d) Dushehari

16. Apple shape cultivar of mango is:
    (a) Rosica
    (b) Rumani
    (c) Totapuri
    (d) Tomy Atkin

17. Suitable temperature for banana cultivation is:
    (a) 17-20°C
    (b) 20-30°C
    (c) 30-34°C
    (d) 8-15°C

18. Amritsagar is important cultivar of:
    (a) Mango
    (b) Bael
    (c) Banana
    (d) Guava

19. Papaya is commercially propagated by:
    (a) Softwood cutting
    (b) Seeds
    (c) Patch budding
    (d) Grafting

20. Enzyme present in pineapple is:
    (a) Papain
    (b) Caprine
    (c) Bromelin
    (d) Bromacil

21. **Adam's Apple** is important rootstock of:
    (a) Apple
    (b) Fig
    (c) Sapota
    (d) Almond

22. Jackfruit is well acclimatized in which type of climate
    (a) Temperate
    (b) Tropical
    (c) Sub-tropical
    (d) Arid zone

23. Carambola is native of:
    (a) Japan
    (b) Indonesia
    (c) Thailand
    (d) China

24. Chemical used for control for pre- harvest drop of citrus fruits is:
    (a) 2-4 D @ 10 ppm
    (b) 2-4 D @ 200 ppm
    (c) NAA @ 50 ppm
    (d) NAA @ 500 ppm

25. Star Ruby is popular & commercial cultivar of:

    (a) Pomegranate **(b) Grape fruit**

    (c) Strawberry (d) Peach

□□□

NIPA Publications on Question Bank

S.No	Title	Author	ISBN	Year
1	Agricultural Statistics: A Guide for Competitive Examinations	Kushwaha, K.S.	9789381450314	2012
2	Objective Agriculture Statistics	Kushwaha, K.S.	9789380235615	2011
3	Objective Agronomy	Vishwakarma A.	9789380235127	2010
4	Objective Biochemistry	Sengar R.S	9789381450376	2014
5	Objectives Genetics	Thirugnana Kumar S. & A. Anandan	9789383305605	2015
6	Objective Horticulture	Verma, Anil	9789381450239	2012
7	Objective Microbiology	Nandi, S.	9789381450017	2011
8	Question Bank Agricultural Engineering	Godara, Er. Amandeep	9789385516122	2016
9	Question Bank in Extension Education	Mohammad, Asif & K. Ponnusamy	9788193014493	2015
10	Question Bank in Postharvest Terminology	Gulera, SPS & Anil Kumar Verma	9789380235042	2010